AF815571

ideaFest

Interdisciplinary Journal of Creative Works and Research

Humboldt State University

Volume 2 | 2018

Humboldt State University Press
Humboldt State University Library
1 Harpst Street
Arcata, CA 95521-8299
digitalcommons.humboldt.edu

ISBN-13: 978-1-947112-07-0

Photo Credits of Full Page Images

Cover Page – "Silvics 331"
Climbing in a redwood grove on campus to collect data.
Taken 14 Apr 2018; © Humboldt State University

Anthropology Section Cover (p. 6) – "Soda fired porcelain"
Meredith Smith artwork titled "And Then Run?" soda fired porcelain is now part of the HSU Permanent Art Collection.
Taken 11 May 2016; © Humboldt State University

Astronomy Section Cover (p. 19) – "Starry night"
Humboldt Redwoods State Park is one hour south of campus and a super spot to view the stars beyond the redwoods.
Taken 12 Oct 2014; © Humboldt State University

Mathematics Section Cover (p. 27) – "100%"
Simeon Haynes, an engineering student that I have known for a few years caught up with me today and showed me his most recent engineering exam which he got a 100% on! I thought it would be fun to do a composite for him of his picture and the exam. I often run into him all around campus and if you come for a tour, he just might be leading it!
Taken 27 Sep 2011; © Humboldt State University

Natural Sciences Section Cover (p. 33) – "GSP 330"
In GSP 330, the students learn to measure elevation at the Ma-le'l Dunes just 4.5 miles from campus.
Taken 7 Feb 2018; © Humboldt State University

Back Cover – "What a view!"
HSU is surrounded by the Arcata Community Forest. You can just see the fifth floor of the BSS building.
Taken 24 Apr 2015; © Humboldt State University

CONTENTS

VOLUME 2 | 2018

Acknowledgements

Producing a journal takes a team and I can't thank enough the group of dedicated individuals that helped put this transformative journal together.

Thank you to **Kacie Flynn and HSU Sponsored Programs Foundation** for your incredible support of student research on campus. You are the engine that makes ideaFest the crown jewel in empowering students and celebrating their achievements with the campus and the world.

Thank you to Managing Editor **Carolyn Delevich**. You not only put together a campus research journal, you put together one with content and design to rival any in the world. A publication like this simply does not happen without your high ethics, knowledge, skill, and dedication. The campus owes you a debt of appreciation.

Thank you **Ashley Schumann** for your early work in organizing the journal, editing, and managing peer review. You are greatly missed in the Scholarly Communications office.

Thank you **Dean Cyril Oberlander** for your vision and dedication to celebrate original student research on campus. People see all the big things you do to support student success, but only because of all the little things you do, but never get credit for, is an endeavor like this possible.

Thank you to **Professor Janelle Adsit** for overseeing the first volume and laying the massive foundation upon which this volume stands. You provided all the guidance and support that made this second edition possible.

Thank you to **Professor Adsit's ENGL 422 Advanced Research Writing** students for providing peer review analyses on all submitted manuscripts. The students provided insightful reviews in the name of campus research, and themselves received a unique hands-on experience into this core element of academic research writing.

Thank you to HSU Library's **Garrett Purchio, Cheryl Conner, Kelly Hangauer,** and **Reynaldo Farfan** for providing secondary reviews. You are the best colleagues and a true service to students and student learning on campus.

And a special thank you to all those who put yourself on the line by submitting an article. Whether published or not, you are all risk takers who strive to create change by sharing your voice with the world. Those who were not chosen for this issue, I expect to see your submissions for the next publication. Greatness does not come easy, but once achieved, it is something to cherish.

Kyle Morgan
Scholarly Communications Librarian
Humboldt State University Press

Note from the Editor

As a biologist, I've seen many sides of academic research – from the gritty field work to the final write-up, and ultimately publication, of a piece of work. I am eternally grateful to have had the opportunity to experience yet another side of academic research as the managing editor of this journal. It is such an invaluable resource that we have as a community at HSU – there are no publication costs, the articles are accessible to anyone online for free, and it allows authors to have more control over how their research is presented. In present times, it may seem difficult to feel that our voice is being heard and that the research that we devote our time and energy to matters. However, I guarantee you that your work matters and that staying silent is much worse than the fear of being ignored.

It took many people to put together the final product that we share with you here. First, I would like to thank Kyle Morgan, my supervisor at the Scholarly Communications office. He put a lot of faith and trust in the idea that I knew what I was doing (even when I didn't) and really allowed me the freedom to piece together this issue in my own vision. It has been quite the journey from a researcher trying to get published to an editor, determining who and what gets published. The right balance of guidance and freedom from him made the task much less daunting.

I would also like to thank all of those mentioned in the preceding Acknowledgements page, whose work preceded mine and made my own contribution possible.

Finally, I want to thank all of the authors of this current issue of ideaFest Journal for their courage and determination in getting their work published. I want to thank them especially for their patience and hard work – I realize that as an editor, I can be demanding and maybe even pedantic at times but I hope that you all grew with me as writers and communicators. And for those who submitted articles but were not published, I look forward to the opportunity to work with you again for the next volume to ensure that your voice is heard.

I hope that through the publication of this second volume, the HSU community will be moved and inspired to submit their own work in the coming years. Education should not be squandered, research should not go unpublished, and important and innovative questions in academia should not go unanswered. I encourage all of you reading this current volume to ask yourself what you can do to contribute to the next issue and become advocates of research that our world so greatly needs today.

Carolyn Delevich
Managing Editor, ideaFest Journal
Humboldt State University Press

ANTHROPOLOGY

A Feminist Interpretation of Women's Work with *Koloa* in the Tongan Community

Mele'ana 'Akolo[1*]

[1]Master's Student, Department of Anthropology, Humboldt State University, Arcata, CA
*author correspondence: mka217@humboldt.edu

Introduction

Born in the United States to parents who migrated from the Tonga Islands in the South Pacific, and being raised deeply immersed in a community enriched with Tongan culture, I based my Master's thesis study on the Tongan community in the San Francisco Bay Area. The three main subjects covered for the research were women, *koloa*, and feminism. According to the Tongan Dictionary by C. Maxwell Churchward, *koloa* means "goods, wealth, riches, possessions; what one values" (1959, 270). Anthropologist Adrienne Kaeppler concluded from her research done in Tonga that the most valued and powerful objects culturally for Tongans are pieces of *koloa* (Kaeppler 1999). I identify as a feminist native anthropologist because I am a feminist who is a native member of the Tongan community that I have studied. It is my strong background with *koloa* that has given me the understanding that *koloa* represent Tongan culture. *Koloa* are a variety of cultural materials such as finely woven mats and *tapa* cloth — cloth made from the paper mulberry tree bark. *Koloa* are used for an array of cultural and social events. At birthdays, weddings, as well as church events, *koloa* are used for decorating venues, are worn by community members in the form of a *ta'ovala* (a finely woven mat worn around the waist), and given as gifts (for example large *tapa* cloth) during traditional gift-giving ceremonies (**Fig. 1**, **Fig. 3**, **Fig. 4**). *Koloa* are the wealth, valuables, products, and possessions of women. Women govern *koloa* and the activities, customs, and traditions associated with the *koloa* tradition. A major goal of the research was to add to the existing research on *koloa* by showing how women have contributed to their communities and society through their work with *koloa* from the narratives of women in the Tongan diaspora community of the San Francisco Bay Area. The research was also meaningful to me because I shared my own narrative and experiences as a member of the community with the women. We expressed the same dedication in asserting through the research that our work and contributions with *koloa* are extremely valuable to Tongan society.

Figure 1. Three women wearing a type of *koloa* known as a *ta'ovala* - finely woven mat worn around the waist.

Women produce *koloa* in a variety of ways. They may produce traditional *koloa* in Tonga by weaving fine mats with fine strips made from the Pandanus tree through a long process or produce a large *tapa* cloth, also known as *ngatu ngatu*, which requires the

Figure 2. Different types of *koloa* used to decorate a graduation party. On the wall are *ngatu pepa* and a *fala pāongo.*

Figure 3. Young boy wearing a *ta'ovala* for a special church occasion known as *Fakame* or White Sunday.

work of several women (**Fig. 7**). Women may also sew or knit quilts known as *monomono*, and are considered nontraditional pieces of *koloa* in their Bay Area homes (**Fig. 5**). Focusing on the work of women with *koloa* in my community, I was able to extrapolate the values found with women's work and explain the significance of these values to Tongan cultural continuity in the Bay Area. In developing a Tongan Pacific Islander feminist thought, I utilized feminist theories to describe the magnitude of *koloa* to women and their roles. I found validation of the persistence of values women have around *koloa*, primarily with the dedication women have toward their work that has lasted for hundreds of years.

My research methodology consisted of autoethnography in the Tongan community, where as a member of the community and a woman I conducted an ethnography from a subjective position. I utilized research methods such as observations when attending different cultural events that required the use of *koloa* and conducting interviews with women from the community that own, use, or make *koloa*.

A demographic questionnaire was administered to the women and I engaged with them in participant observations. My goal for the research was to provide deeper meanings for the values of women and their work, using the narratives of women in the community in order to describe *koloa* as a tradition rather than a product. As a feminist native anthropologist my goal was to shed light on where and how women's work substantially factors into Tongan cultural values being preserved and highly regarded, even in places such as the Bay Area where dominant cultural values prevail.

Figure 4. Tongan couple wearing a *ta'ovala* showing their Tongan cultural identity.

Methods

I used several types of research methods to draw out data from the women. I gathered personal memories of *koloa* through interviews, made observations at events where *koloa* were present, and partook in participant observations, helping women busy with their *koloa* activities. I also used methodologies such as autoethnography, a reflexive and feminist approach, feminist analysis, snowball and volunteer sampling, and maximum variation sampling. The population of women I chose to research came from the community in which I grew up.

The two sampling methods I used to gain participants were "snowballing and volunteer" and "maximum variation sampling" (Seale 2012, 145). I used the snowballing and volunteer method by asking women I already interviewed if they knew of any other women

Figure 5. Different types of *koloa* used to decorate a birthday party. The quilt on the wall is a *monomono*, which is a quilt sewn by women made from Western materials and classified as a Westernized type of *koloa* or *me'a fakapalangi.*

who would be interested in participating in the research. Participants also voluntarily suggested other women to interview. This method was not effective for gaining participants. Instead, asking women at events to participate in the study was more effective. There were typically larger crowds of people at an event and I was able to get several women to participate by informing them about the study, requesting their participation, and, if they accepted, getting their contact information to set up an interview at a later date.

To develop a variety of women with different backgrounds to participate in the study I used the maximum variation sampling method. This method was crucial for obtaining participants because I planned to have a small sample size of women, but I wanted a diverse group. At events, I asked women personally if they would be interested in letting me interview them. I also contacted women through social media sites such as Facebook, called or emailed women I knew, and asked women I interviewed if they could put me into contact with other women who would be interested in contributing to the study. Building a diverse group of women to pull information from was necessary to validate my findings. I managed to include women who were born in Tonga, born in the United

States, were experienced with *koloa*, had little experience, and established a wide age range of women. Having extremes within the group provided diversity. The women were also from various religious backgrounds, cities, and had varying levels of experience with *koloa*. This was to produce a diverse set of data to use for descriptions (Seale 2012). The interviews were strictly on a voluntary basis and no compensation was given for participation.

All of the interviews were one-on-one. They were conducted in several locations and for all of the interviews, arrangements were made to meet at a specific time in a private or public location where the woman felt comfortable. The interviews took place in the homes of some of the women, in my home, at a church, in coffee shops, restaurants, parks, and libraries. The interviews took at least 45 minutes, and the longest interview I conducted lasted nearly three hours. On average, the interviews lasted one hour. They were all recorded using a voice recording device. The women were given a hard copy demographic survey questionnaire to fill out before the interviews, which I used to gather basic information such as age, where they were born, their employment status and if they had children.

Figure 6. Young girl (center) is wearing a *teunga tau'olunga* or traditional Tongan dancing costume.

I attended five events in the Bay Area. The events took place at churches, a home, and a hall. Women active in their communities engage in events that involve *koloa* activities. A birthday I attended was decorated with traditional and nontraditional *koloa*. There was a gift-giving ceremony where *koloa* were given to honored guests. I participated in two events that I observed. One event was organized to promote the sales of traditional *koloa* pieces manufactured by a group of weavers specializing in making *koloa*. I worked with a woman representing the group, who brought the *koloa* from Tonga to be sold at an event we organized held in a church hall. **Figure 8** is a photo of myself at the bazaar of *koloa* products on display.

I used a feminist approach to this research by choosing only to include the participation of women to give prominence to their narratives, forming a platform for women to be active agents in creating information about *koloa*, Tongan culture, and their significant contributions to cultural continuity in their community. This type of involvement by women in the community had not been substantially documented in previous research, which also lacked a feminist native anthropologist's perspective. The criteria for participants were that they were Tongan and women. Their level of expertise with *koloa* was not my main concern since I wanted a diverse group of women. However, the women were required to have some knowledge of *koloa*. My interview questions were structured to obtain information from the women about how their womanhood was defined by *koloa* as well as how they valued *koloa*. The questions were directed toward the women to initiate answers that led into stories of experience, how their experiences influenced them, how or in what ways they valued *koloa*, and how much their values reflected their relationship with women in their families and the community. The questions also touched on topics around gendered roles, identity, womanhood, and the work that they did with *koloa*.

The point of using the reflexive approach was to

highlight the narratives of the women, making their stories about their experiences with *koloa* essential for understanding the value of women's work. A reflexive approach engages participants in the study to let their interpretations of culture, customs, and traditions be integrated into the development of accounts or descriptions of the particular culture under investigation. Their contributions to the study are emphasized where they become a focus of the study. The autoethnography complemented the reflexive approach in that I incorporated my own narrative of my experience with *koloa* rather than the research being conducted by an outsider. As a Tongan woman, it was important for me to be honest about the connection I had with my subject matter. As a subject myself, I was able to make observations and gather data that were not only meaningful to my participants but also meaningful to me, even before analysis. Placing myself beside my participants allowed for theorizing about the women from a place where we had a similar social and cultural history, lived experiences, and similar social constructions of our womanhood (Bolles 2001, 35). My ambition was to construct theories about what to study with *koloa* and women, so that information being obtained would reflect the multiple interrelated oppressions that women face, how they combat them, and how these actions were a part of their role as women. I also wanted to bring awareness to the contributions of women and their *koloa* in the community (McClaurin 2001, 62). The reflexive approach allowed for explaining the pragmatic qualities of *koloa* from the women's experiences where women viewed *koloa* as undeniably representative of their cultural identity. I also shared this same view as the women and was instrumental in highlighting this because of my subjectivity with the research. For example, *koloa* have always been a way for me to express my cultural identity. They have been a tradition historically controlled by women and they represent aspects of my womanhood (Seale 2012). I took advantage of my position politically as a feminist activist by challenging as well as altering common traditional practices with research methods used in the field by studying women from my own native

Figure 7. Traditional *ngatu ngatu* type *koloa* (folded piece on top of fine mat) given as a gift during gift-giving ceremony. It is folded and placed on top of the fine mat type *fala fihu*, shown here, for presentation.

community. This was to underline the point that anthropology has changed from its historic past of colonialism, where subjects' contributions were not noted and researchers were typically outsiders (McClaurin 2001, 62). As a feminist it was paramount to develop descriptions of the *koloa* tradition, arguing for the significant value of women's work and declaring their contributions with *koloa* as fundamental to maintaining traditional Tongan culture. As a native this was imperative for me to do because Tongan cultural identity was extremely valued by the women, as well as the community. *Koloa* are things I grew up with and have grown to value as a woman.

The use of autoethnography permitted me to stylistically textualize my experiences, along with those of my subjects, through a theoretical lens where we could interpret, describe, analyze, and develop accounts of the *koloa* tradition as a group (McClaurin 2001, 64). Autoethnography allowed me to have "transformative ethnographic knowledge production" that countered the sometimes "frozen/static ethnographic represen-

Figure 8. Display of a variety of different *koloa* for sale held at a church hall in Redwood City, California.

tations" of traditional ethnography (McClaurin 2001). Traditional ethnography is still valued due to its holistic nature and attention to detail. However, it has been found to have flaws because of biases the researcher had. As an outsider the researcher may judge the culture of those they are studying based on the standards of the researcher's own culture (McClaurin 2001, 64). The benefit of autoethnography was that I could attempt to fully interpret and theorize about the women, their community, and *koloa* based on how I engaged not only as a feminist, but as a native anthropologist standing together with the women in the Tongan community. Autoethnography is, as Irma McClaurin describes, "blending the grounded, detailed descriptions that come from ethnography with the poetics of autobiography to create autoethnography," which offers anthropologists like myself the opportunity to express their connectedness with the subjects they are studying by discussing experiences collectively as natives (McClaurin 2001, 71).

A feminist analysis was employed for the research because grounding descriptions in feminist theories was necessary to truly understand the values of *koloa* and women's work, beyond the values of wealth or economic contributions. Feminist theories that were a focal point for the analysis included gender with regard to division of labor, agency as a form of self-definition, women working in the globalized world, Black feminist thought where women of color resist assimilating to dominant culture, and postmodernism with respect to subjectivity. Theories related to gender were pertinent to the analysis because they emphasized that division of labor with *koloa* responsibilities is fundamental for women to maintain their domain with *koloa* to continue the tradition and promote enculturation. Theories related to women as agents defining their womanhood called to attention the responsibilities women have with *koloa* that provide relevant value to their gender roles and the division of labor between men and women when it comes to the *koloa* tradition. Women have an array of responsibilities with *koloa* that men do not have, that define who they are as contributors to the community and society at large. It is women who have governed this tradition historically which has served as a mechanism for women to define themselves, but more importantly to define aspects of Tongan cultural identity. They must possess *koloa* because their mothers and their grandmothers possessed *koloa*. They must be knowledgeable on how to care for their *koloa* to keep them from getting damaged. They must know what the different types are, their names, and values. They must know how to present their *koloa* during gift-giving ceremonies for all sorts of social and cultural occasions. They must provide *koloa* to their family, church, and community members to wear for social and cultural events, to present as gifts for a family member's wedding or funeral, as well as for decorating a traditional Tongan event. They are responsible for providing group members with cultural materials necessary for carrying out Tongan customs and traditions highly valued in Tongan society. Tongan culture is constructed out of longstanding customs and traditions, one of which are *koloa*.

As self-defining agents of their womanhood, women have valued their work with *koloa* to the point that the tradition has survived throughout the years, giving women's work authority and power within the Tongan community. Theories related to Black feminist thought, particularly with resisting assimilation to dominant culture, assert that maintaining *koloa* in the Tongan diaspora community is a form of resistance to dominant culture (Hill-Collins 2000). Women continue to uphold their responsibilities not only in present time, but also in places all over the world far removed from the Tonga Islands. They are women doing women's work in a globalized world. Women sell and trade *koloa* on social media sites like Facebook and ship their *koloa* from Tonga to places such as the United States, Europe, Australia, New Zealand, and Asia. By taking a feminist approach to my research from a subjective standpoint I diverged from traditional research practices in the discipline to embrace new methodology. I am promoting change with conducting research and presenting information about Tongan culture that distinguishes women's contributions economically, politically, culturally, and socially from other members of society.

The feminist analysis was directed toward covering womanhood to allow the development of a Tongan Pacific Islander feminist thought on women's work with *koloa*. Information obtained from the women and information acquired from the literature review were carefully evaluated to locate areas where relevant contributions were made, addressing or relating to topics such as womanhood, culture, *koloa*, customs, tradition, and men. That information was then grounded in feminist theories, anthropological theories, and finally, feminist anthropological theories. Theories in feminist anthropology that were applicable to the analysis focused on contributions that women have made to culture that have not been documented or attributed to women around *koloa* and cultural continuity. Feminist anthropological theory utilized in the analysis focused on providing accounts of women's work not associated with their reproductive role or domestic work. The literature review was an extensive process that involved identifying pertinent literature as well as identifying theories in which to ground the descriptions.

Literature Review

Contributing to the existing research on Tongan diaspora, women and *koloa*, feminism, and cultural identity required diverging from the path paved by previous researchers and situating my research within the realm of a feminist native anthropologist. Positioning my work as feminist involved reviewing feminism both in and out of anthropology, grounding my research in feminist theories reflecting the experiences of women of color. Developing a Tongan Pacific Islander feminist thought involved identifying what women did to counter forms of oppression, such as assimilation (doing away with one's native culture to adopt or conform to dominate cultural values). By building on existing theories, analysis of the research focused on the ways women resisted inequalities with race, class, and gender by identifying values women had developed around *koloa* and women's work that defined womanhood. For example, I analyzed the obligations women identified as valuable in maintaining cultural identity with their work with *koloa* that defined their womanhood. I analyzed the ways the women overcame the strain of carrying such huge responsibilities in order to maintain the *koloa* tradition and women's status with *koloa*. To what extent did internalizing the burdens with their responsibilities factor into the structuring of positive ideals around gendered division of labor, cultural identity, social as well as hierarchical status, and womanhood?

Current theories and descriptions of women and *koloa* focus on production, the economic value of *koloa*, and women controlling *koloa*, with less attention paid to *koloa* as a tradition and work women do as cultural providers. It was crucial that I contribute to these theories from a feminist native perspective. My goal was to emphasize the changes taking place with anthropological research from the past that differed from my approach, in which the subjects' narratives were the main focus. I searched for literature on cultural identity and Tongan diaspora communities that did not make significant reference to the use of *koloa* in showing cultural identity. My goal was to develop descriptions, accounts, and theories based on narratives from women explaining why or how *koloa* are valuable to Tongans, particularly Tongan women culturally identifying as Tongan.

Helen Morton Lee discussed Tongan cultural continuity in Tongan diaspora communities and, when discussing Tongan identity, she claimed that there were certain qualities of this identity that people in her study strived to live up to and continue (Lee 2003). Lee emphasized the importance of Tongan cultural identity to Tongan people living in diaspora communities. Developing narratives in her research from women in the community around cultural identity would have allowed women the opportunity to raise the topic of the *koloa* tradition in their conversations as a way to express one's Tongan cultural identity, as they are fundamental aspects of Tongan culture and necessary for the culture to exist.

Ping-Ann Addo and Phyllis S. Herda described traditional cultural practices of *koloa* production in Tonga of one particular type known as *ngatu ngatu*, a traditional *tapa* cloth made from the inner bark of the paper mulberry tree. Their discussion on *ngatu ngatu* production was a common theme found in existing literature on *koloa*, where *ngatu ngatu* was presented repeatedly as if it were an ideal example of a type of *koloa*. The theme developed in the literature referenced the production of *ngatu ngatu* to describe the labor involved with producing *koloa*. Discourse in the literature specifies the process of making this type of *koloa* to address the presence of a division of labor between men and women with *koloa*. The reoccurring referencing of *ngatu ngatu* production alluded to the concept that *koloa* are mainly products or material things that put excessive labor on women and not men, which is a misrepresentation of the true value of *koloa*. *Koloa*, as a tradition, are fundamental to continuing Tongan culture and are completely maintained by women. With my own research, the goal was to stress that women controlled more than just the production of *koloa* through traditional or nontraditional practices. They encountered conditions for holding their leading position with *koloa* that demanded an extensive amount of work such as production, but production was not the only condition women encountered in my study. The extensive work involved with *koloa* that women endured outside of production consisted of learning how to use *koloa* for different traditions, caring for one's *koloa* to maintain its condition, and finding ways to accumulate *koloa*. These were responsibilities only women underwent, that men did not have to be subject to. Some of the women did not produce *koloa* at all, but they were a fundamental part of their womanhood that they valued because they learned these values through enculturation. During enculturation the women described the valuable time they spent with other women learning about the *koloa* tradition from their grandmothers, mothers, aunts, sisters, elders, and peers. Seeking to emphasize the diverse obligations women had with *koloa* brought about memories of loving moments with family or friends. They were representatives of knowledge when it came to culture because they became masters of their craft in every way, not just in production or extensive labor.

I turned to literature on feminism in and out of anthropology because the theories developed there were aimed at analyzing the work of women as contributions, a form of power and knowledge. These deeper meanings of the work women do with *koloa* can be overlooked by a researcher not trained in *koloa* activities. My goal was to use my own personal knowledge of *koloa* to touch on cultural meanings and values related to women, *koloa*, and feminism that have been previously overlooked.

Findings

As a Tongan woman growing up in the United States with a mother who possessed and used all types of *koloa* for responsibilities to her family, church, and community, different types of *koloa* were not just a form of wealth, something my mother possessed for the sake of possessing them. They had other meaningful and valuable functions. In spite of *koloa* being primarily defined as a form of wealth by the Tongan Dictionary and previous scholars, I wanted to look into other factors giving value to the different types of *koloa*. I agree that they are valued and powerful as Kaeppler pointed out, but my goal was to find out in what ways were they valued by all women and how that factored into our work perpetuating the tradition over time. My mother taught me about *koloa* starting from a young age. My first memories of *koloa* were the fine mats my mother stored under my mattress. There, they stayed smooth and preserved in presentable condition. For example, the fine mats, or *ta'ovala*, pulled from underneath my mattress would be as crisp as a shirt straight from the dry cleaners, still wrapped in plastic, and ready-to-wear. I learned this practice of preserving my *ta'ovala* through enculturation specifically from my mother. Women in the community contribute to society specifically with their work around *koloa* through enculturation because they not only teach their daughters about *koloa*, but they teach society in general through their publicly displayed practices marked by their domain with *koloa* activities.

I discovered from women in the Tongan community that they held strong values around *koloa*. Their first memories of *koloa* started at a young age; it was something they had been learning about since they were in elementary school. However, they did not realize the full potential of *koloa* because they were too young. Women generally did not internalize the *koloa* tradition as their gift consisting of womanly values that went beyond just cultural values until they had families of their own. Their philosophy around their responsibilities was that they were guardians of *koloa*. The values they held factored into their feelings of obligation caring for, upholding, and preserving the tradition. Women felt keeping with their responsibilities showed they were keeping their integrity as women. Keeping the *koloa* tradition showed love toward one's family, church, community, and culture. This involved having *koloa* at all times in case of an event. The women felt it was shameful if a woman could not fulfill her responsibilities because a woman without *koloa* is a woman without Tongan culture or identity. *Koloa* represented their cultural identity, but it also represented their womanhood and resistance to assimilating to dominant cultural values associated with Westernized notions of womanhood. Women were relied upon by their families, communities, and other women to fulfill responsibilities with *koloa*. There was a sensibility with the women that *koloa* were something Tongan women should have in case there was a wedding, birthday, or event in the family, church, or community.

Women acting as guardians of *koloa* appreciated their position by structuring values to keep *koloa* within their domain, such as in the privacy of their home. Here they could teach family about their values with *koloa*, how they cared for it, used it, and what the different types were. It could also be in the public where they conduct presentations with *koloa* such as the gift-giving ceremony. In their domain, women relied on each other to transmit knowledge and values with customs, traditions, and culture to the community. Their domain with *koloa* activities was something they ruled and over time women have been able to maintain their position within their domain as well as maintain the *koloa* tradition, making it a staple in Tongan culture.

Despite the enveloping effect of dominant culture, women have resisted assimilation by maintaining the *koloa* tradition. However, not all women in the diaspora community uphold this tradition. One woman, referred to as Fa, explained that if a woman was to have an event, but she did not have *koloa*, rumors would circulate about how she chose to conduct her responsibilities. Women can choose to work with *koloa* when fulfilling their responsibilities to the family or the community. When they choose not to, they are accused of not showing their Tongan identity or overly conforming to dominant cultural values, which are not encouraged if those values take away from one's Tongan identity. The practice of women spreading rumors about women not contributing their *koloa* serves as a mechanism of social control discouraging women from doing away with their *koloa* and instead encouraging women to keep with the tradition. *Koloa* was used to show cultural identity or "Tonganness" by women completing their responsibilities with *koloa*. Traditional pieces that were specific to the Tonga Islands held a greater value than nontraditional pieces, even though all types of *koloa* were considered useful for fulfilling customs and traditions. When women broke with tradition it was as if they were *fiepalangi*, literally meaning "trying to be white", doing away with the values that they were brought up with.

Establishing Tongan culture in the Bay Area was crucial to building the Tongan community. Women contributed significantly to building communities because they were knowledgeable when it came to culture and they possessed the resources essential to building the foundations of culture.

Discussion

It was evident from stories of the women in my community that they tremendously valued the *koloa* tradition. Discovering these values the women shared was made possible through seeking their narratives on *koloa* and making them a crucial part of the study. There were several reasons why women valued *koloa*, ranging from responsibilities to the family to self-integrity. One reason I heard repeatedly by the women was that *koloa* represented who they were. *Koloa*, particularly traditional *koloa* pieces imported from Tonga, were representative of Tonga. Tongan culture was what the women wanted to preserve about their identities. The women valued this identity to the point that they committed to their responsibilities with *koloa* for their families, churches, and communities by enduring the labor involved, the time it took to make and preserve *koloa* of all types, and also working with other women to provide *koloa* for big community events. They valued their connection to their role as Tongan women preserving the tradition. Despite the responsibility that came with maintaining *koloa*, women willingly continued to withstand the challenges of their position as guardians to maintain their cultural identity and acculturate members of the community to do the same. One main reason *koloa* has been preserved historically by women overtime is that *koloa* are the wealth of women and they have been defined by women as women's responsibility. The Tongan identity people seek when wearing their *koloa* is definite, but the tradition also entails social and cultural values specific to women over any other member of society because they are the guardians of the *koloa* tradition. Tongan society has accepted women as guardians of *koloa* because they have maintained this culturally-defining tradition for generations. Even though women were respected for their work with *koloa* in the community, they were not given as much recognition for this in the existing research. The stories of the women are what mattered the most for this study because they were the basis for giving women the recognition they deserved for their work with *koloa* that has persisted overtime.

Koloa are symbols of Tongan culture in the Bay Area and in Tonga. They are highly valued by members of society from the young to the old and from commoners to the monarchy. Occasions of different types, from birthdays to school reunions, consist of customs and traditions ceremoniously carried out to complete the events as specific to Tongan cultural values that structure society. Figures 5, 6, and 7 show examples of how women use *koloa* to make their events traditionally Tongan. What makes *koloa* have great value is not primarily economic value. After all, to Tongan people it is no mystery that *koloa* of all types carry some form of economic value. *Koloa* carry significant value because they are a tradition that has become perceptible to society as symbolic of culture because of the work women do. There are a variety of different types of *koloa* that women must not only be knowledgeable of, but also own. Women must know the lengths of these different types of *koloa* and how to present them at their event for the gift-giving ceremony. For some women, this is knowledge that they have learned in school, but all of the women have learned about *koloa* from their upbringings by seeing their mother, grandmothers, aunts, and other women in the community fulfilling their *koloa* responsibilities over their lifetime. *Koloa* are necessary for completing customs and traditions with funerals, church functions, weddings, or birthdays. Without *koloa*, those events would not be traditional Tongan events. *Koloa* represent what epitomizes Tongan culture because they were made from the grandmothers, mothers, daughters, sisters, and wives of the Tongan community. Those representations of culture are what make *koloa* the benchmark for keeping with the complexities of Tongan culture, guarding customs and traditions because the women have guarded this focal point of culture generation after generation. Women's work with *koloa* is relevant because the cultural values that revolve around the *koloa* tradition have impacted society and made known the power of women with Tongan culture.

Conclusion

Choosing to research as a feminist native anthropologist has been eye-opening. I was familiar with the experiences of the women I was studying because their experiences were my experiences, my mother's experiences, and my grandmother's experiences. Utilizing this perspective and supporting change from the slow-changing past of anthropology by making use of new methodologies such as autoethnography and a feminist approach, prove that the discipline is open to new ways of conducting research. This autoethnographic study draws awareness to the women's valuable work and encourages their growth. I used unconventional methodologies such as a feminist approach, native approach, and autoethnography because they were essential for considering how women were paramount when it came to cultural identity. Developing knowledge explaining the caliber of women's work over time through their narratives served as a platform for women to expound what *koloa* means to them. *Koloa* has a special value to me personally because I am a woman and they make me feel proud when I am a witness to the customs and traditions carried out publicly with *koloa* that represent my culture.

I deviated from the path paved by previous researchers by taking focus away from the economic value of *koloa* and emphasizing the strong values women have for fulfilling customs and traditions with *koloa*. The division of labor in society between women and men due to the tradition was valued by the women because the tradition was a way for them to contribute to society in a way that men could not. Describing the *koloa* tradition as controlled by women through the voices of women, shed light on the necessity for the division of labor. This validates the dominant role women have in society because women associate *koloa* with womanhood, where men are impertinent when it comes to *koloa* responsibilities and the work involved. Men historically do not have anything to do with *koloa*, except that they wear certain types such as the *ta'ovala* to express their Tongan identity. In places such as the United States, gaining equality between women and men has sparked social justice movements with groups of people protesting for women. My focus on the lived experiences of women contributing to society with *koloa* supports division between women and men because women can play a powerful role within their families, churches, and community through their work with *koloa*. It establishes Tongan culture for community members to engage in and identify with.

Developing a Tongan and Pacific Islander feminist thought was possible because I put my focus in finding out where women's work formulates knowledge either in the home or in public, as well as among women. I was able to understand the origin of their knowledge through listening to their first memories seeing women preparing all types of *koloa*. Women continue to distribute their knowledge to society as a resource for people to find ways to decolonize their minds by resisting Western cultural values and expressing their Tongan cultural identity. Women have cultural knowledge that is invaluable to society and coming to this conclusion would have only been possible by observing the work of women through a feminist lens. Moreover, taking advantage of my native perspective allowed me to spot the cultural knowledge women gained through their commitments with *koloa* and argue for the relevance of their cultural knowledge to Tongan society. My part in the research as a native anthropologist revealed how the empirical foundations of anthropology have changed to further facilitate women studying women from their own community. A primary goal of doing this type of anthropological work is to show that the possibilities for change are attainable and deviation is needed for innovative anthropological research.

Acknowledgements

I would like to thank the women in my study who shared willingly their experiences with koloa. Thank you to my committee members for all their help. Lastly, I thank my family for all their love and support.

References

Addo, Ping-Ann. "Commoner Tongan Women Authenticate 'Ngatu Pepa' in Auckland." Pacific Arts 3/5, (2007): 60 – 73.

Bolles, A. Lynn. "Seeking The Ancestors. Forging a Black Feminist Tradition in Anthropology." In Black Feminist Anthropology. Theory, Politics, Praxis, and Poetics, edited by Irma McClaurin, 24 – 48. New Brunswick, New Jersey and London: Rutgers University Press, 2001.

Churchward, C. Maxwell. Tongan Dictionary. Tonga: Government Printing Press, 1959.

Herda, Phyllis S. "The Changing Texture of Textiles in Tonga." The Journal of the Polynesian Society 108, no. 2 (1999): 149 – 167.

Hill-Collins, Patricia, Black Feminist Thought, Knowledge, Consciousness, and the Politics of Empowerment, Second Edition. New York and London: Routledge, 2000.

Kaeppler, Adrienne. "Kie Hingoa: Mats of Power, Rank, Prestige and History." The Journal of the Polynesian Society 108, no. 2 (1999): 168 – 232.

Lee, Helen Morton. Tongans Overseas, Between Two Shores. Honolulu: University of Hawai'i Press, 2003.

McClaurin, Irma. "Theorizing a Black Feminist Self in Anthropology. Toward an Autoethnography Approach." In Black Feminist Anthropology. Theory, Politics, Praxis, and Poetics, edited by Irma McClaurin, 49 – 76. New Brunswick, New Jersey and London: Rutgers University Press, 2001.

Rodriguez, Cheryl. "A Homegirl Goes Home. Black Feminism and the Lure of Native Anthropology." In Black Feminist Anthropology. Theory, Politics, Praxis, and Poetics, edited by Irma McClaurin, 233 – 257. New Brunswick, New Jersey and London: Rutgers University Press, 2001.

Seale, Clive. "Sampling." In Researching Society and Culture, edited by Clive Seale, 134 – 152. London, Thousand Oaks, New Delhi, and Singapore: Sage, 2012.

Small, Cathy A. "The Birth and Growth of a Polynesian Women's Exchange Network." Oceania 65, (1995): 234 – 256.

Teilhet-Fisk, Jehanne H. "To Beat Or Not To Beat, That Is The Question: A Study On Acculturation And Change In An Art-Making Process And Its Relation To Gender Structures." Pacific Studies 14, no. 3 (1991): 41- 68.

Weiner, Annette B., Women of Value, Men of Renown, New Perspectives in Trobriand Exchange. Austin and London: University of Texas Press, 1976.

Weiner, Annette B. "Why Cloth? Wealth, Gender, and Power in Oceania." In Cloth and The Human Experience, edited by Annette B. Weiner and Jane Schneider, 33 – 72. Washington and London: Smithsonian Institution Press, 1989.

Weiner, Annette B. Inalienable Possessions. Berkeley, Los Angeles and Oxford: University of California Press, 1992.

ASTRONOMY

Searching for a Connection Between Radio Emission and UV/Optical Absorption in Quasars

Sean Haas[1*], Carla Quintero[2], Paola Rodriguez Hidalgo[3]

[1]Undergraduate Student, Department of Physics and Astronomy, Humboldt State University, Arcata, CA
[2]Undergraduate Student, Department of Physics and Astronomy, Humboldt State University, Arcata, CA
[3]Associate Professor, Department of Physics and Astronomy, Humboldt State University, Arcata, CA
[*]author correspondence: ssh239@humboldt.edu

Introduction

Characteristics of Quasars

Quasars are among the most distinctive and most energetic objects in the night sky. Powered by accretion onto an accretion disk around the central supermassive black hole, quasars emit huge amounts of radiation over the entire electromagnetic spectrum. Quasars are also very distant, making them difficult to study.

Spectral observations are the main tool used to study quasars, due to this combination of distance and strong emissions [1]. The characteristics of observed quasar spectra differ greatly from the spectra of stellar bodies. Quasar emissions cover a large range of wavelengths instead of resembling a discrete blackbody curve, as observed in stellar spectra. Additionally, quasar spectra are highly variable, especially in the UV/optical region. That is to say, a quasar's spectrum can change dramatically over time [2].

Outflows and Jets

One of the more prominent features of quasars are axial radio jets, which are present in roughly 10% of quasars. These jets, composed of highly accelerated charged particles, emit large amounts of radio waves [2]. These radio emissions make it possible to detect quasar jets by simply measuring the object's radio flux. Quasar outflows are another more recently documented phenomenon of note. An outflow is an event where a quasar ejects large amounts of matter. These outflows can be observed in UV/optical spectra as blueshifted absorption features. This blueshift occurs as matter ejected from the quasar travels towards Earth while absorbing light from the quasar. One problem with this observation method is that only outflows positioned between the quasar and Earth, and traveling towards the Earth, can be detected. Another spectral signature of these outflows is that their absorption features tend to appear as broad absorption lines (BALs) rather than thin absorption lines. This broadening of absorption lines is due to the matter in the outflow moving at differing speeds. The speed of these outflows can be measured by comparing the location of the BAL to its expected location in the quasar's rest frame. Outflows with higher observed velocities, $v > 0.1c$ (10% the speed of light), are termed extremely high velocity outflows (EHVO) [3].

Usage of CIV

The absorption of CIV, triple-ionized carbon, is one of the most commonly used markers for these observations. This ion is chosen for a few reasons. Firstly, carbon is a very common element, so it can be assumed that any given outflow will contain at least some amount of carbon. Secondly, CIV absorption features are easy to observe, as they fall into the UV/optical region of the electromagnetic spectrum. This combination of factors makes CIV a useful marker for studying outflows in quasars.

Methods

Two main data sources were used in this study: the SDSS DR9 and VLA FIRST. SDSS is an automate UV/optical survey of the night sky. Data from this survey are periodically published as numbered data releases and are made available online in the form of optical imagery and spectrographic data. SDSS DR9 was used as a source of spectra for the spectral analysis program. VLA FIRST is a radiometric survey of the night sky that targets faint objects and was chosen for this study because it contains observations for many of the objects surveyed in SDSS. Cross correlation used the DR9 Quasar Catalog, a value-added catalog of quasar data that contains information from SDSS DR9 and VLA FIRST [4].

Inputs for the spectrum analysis program were chosen using a number of criteria. One consideration was signal to noise ratio (S/N). Samples with $S/N \geq 10$ were chosen because BAL proved to be hard to detect over noise in spectra with lower S/N. Another consideration was redshift (z). Redshift is a measurement of how fast something is moving away from us which is observed by measuring how the Doppler effect shifts the entire spectrum. A redshift greater than one means the entire spectrum will be shifted towards larger wavelengths. Due to the expansion of the universe we can use redshift to tell how far away an object is. CIV was used as a marker for absorption, as it is a common ion in quasar outflows and, therefore, produces strong absorption features. Since this study investigated outflows with $0.1c < v < 0.2c$, CIV features would appear at around 1200 Å in the rest frame of the spectra. Due to this, quasars with $z \geq 1.9$ were chosen, as this redshift moves the CIV features to around 3600 Å, within the range of DR9's coverage wavelengths [4]. The SDSS DR9 catalog contains 87 822 quasar samples. After applying these filtering criteria, $S/N \geq 10$ and $z \geq 1.9$, the sample was reduced to 6760 quasars.

The first step in the study was to find a sample set of quasars with EHVO so that their radio properties could be examined. A Python program was written to search for quasars with EHVO. These quasars were found by searching quasar spectra from the SDSS DR9 quasar catalog for broad CIV absorption features between the SiIV emission line and the Lyman alpha forest. This region was used because it can reasonably be assumed that any broad absorption features in the region are blueshifted CIV absorption features, and, therefore, evidence of outflows. This search was accomplished by first applying a powerlaw fit to normalize the spectra's continuum. After normalization a three point median boxcar method was applied to smooth the continuum. Next, the program identified quasars with possible EHVO by searching for BALs by integrating between the powerlaw fit and the continuum. Using this process, the program was able to generate a list of quasars with possible evidence of EHVO. The quasars that this program identified as having evidence of EHVO were then visually inspected. From this inspection a set of quasars with definite evidence of EHVO was selected. These samples were then cross correlated with VLA FIRST to identify their radio properties. This cross correlation was car-

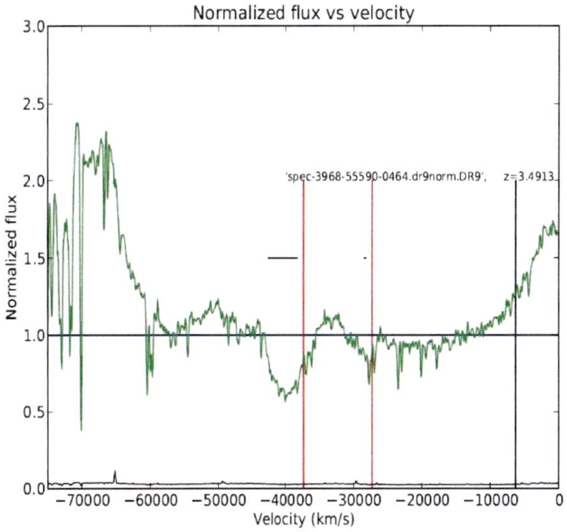

Figure 1. Example of normalized quasar spectra from our final data set. SDSS spectra (green) smoothed and plotted with a guide line at 1 to aid visual inspection (blue), CIV absorption found outflowing at speeds between 0.1c and 0.2c (red), and where CIV would be if the absorption marked with a red line were SiIV absorption instead (black). The feature at -40 000 km/s is a BAL corresponding to CIV, indicative of an outflow.

ried out using the DR9 Quasar Catalog [4]. An example of a spectrum with evidence of EHVO found by this program is shown in **Figure 1**.

Conclusions

Results of Spectral Analysis

From the filtered set of 6760 spectra from the SDSS DR9 catalog, the analysis program identified 41 spectra as having possible evidence of EHVO. Visual inspection was carried out to remove obvious non-EHVO spectra and reduced this set to 23 quasars with definite evidence of EHVO. Cross correlation with VLA FIRST radiometric observations was carried out using data in the SDSS DR9 quasar catalog. From this cross correlation it was found that none of the quasars in the final sample set had any measured radio flux.

Conclusions

From these results, it is evident that radio emissions are not a prerequisite for EHVO in quasars. Additionally, since axial jets are strong radio emitters, this evidence suggests that jets are not a prerequisite for EHVO in quasars.

SDSS Targeting

A total of 3076 quasars in the SDSS DR9 quasar catalog, about 3.5% of samples, have some radio flux. However, this sample may not be completely representative of the general population of quasars. Originally, SDSS surveys targeted quasars based on their radio properties. Specifically, some early targets in the SDSS-I/II were selected due to their radio emissions as measured in VLA FIRST [5]. Due to this targeting there may be a disproportionately large amount of quasars with measurable radio flux in SDSS sample sets than would be found in the general population of quasars. This sampling bias needs to be considered when using SDSS data sets.

Acknowledgements

We would like to acknowledge our advisor, Dr. Paola Rodriguez Hidalgo, as well as Abdul Moiz Khatri and Pat Hall.

References

1. Peterson, B. M. (2008). The central black hole and relationships with the host galaxy. New Astronomy Reviews 52: 208 – 252.
2. Netzer, H. (2013). The physics and evolution of active galactic nuclei. Cambridge: Cambridge University Press.
3. Hildago, P. R., Hamann, F., Nestor, D., Shields, J. (2007). High velocity outflows in quasars. [Presentation]. Paper presented at the ASP Conference Series.
4. Pâris, I., *et al.* (2012). The Sloan Digital Sky Survey quasar catalog: ninth data release. Astronomy & Astrophysics 548. doi:10.1051/0004-6361/201220142.
5. Ross, N. P., et al. (2012). The SDSS-III Baryon Oscillation Spectroscopic Survey: quasar target selection for data release nine. The Astrophysical Journal Supplement Series 199 (3): 1 – 29.

How Many Quasars Have Extremely High Velocity Outflows?

Carla Quintero[1*], Sean Haas[2], Paola Rodriguez Hidalgo[3]

[1]Undergraduate Student, Department of Physics and Astronomy, Humboldt State University, Arcata, CA
[2]Undergraduate Student, Department of Physics and Astronomy, Humboldt State University, Arcata, CA
[3]Associate Professor, Department of Physics and Astronomy, Humboldt State University, Arcata, CA
[*]author correspondence: cpq11@humboldt.edu

Abstract

Quasars are luminous celestial objects that reside in the core of the most massive galaxies, where a supermassive black hole resides, powering the quasar's system. In some cases, quasars have been observed emitting outflows – matter that is expelled out of the quasar's environment instead of falling into the black hole. Our study's purpose was to determine: (1) how many quasar outflows have speeds larger than 10% of the speed of light (c) and (2) whether there was any correlation between outflow and radio emissions from quasars. Radio emissions are produced by quasars jets, which are another piece of the system but are present only in 10% of quasars [1]. To answer these scientific questions, our group analyzed 6760 quasar spectra from the Sloan Digital Sky Survey Data Release 9 (SDSS DR9), a public data set [2], searching for those outflows with speeds larger than 0.1c and cross-correlating with the corresponding radio information. We have selected the brightest quasars and the ones at the right distance from Earth to be able to study the potential presence of these outflows. This entails that the quasar outflow is in between the quasar and Earth, with the outflow traveling towards Earth. In this paper we discuss the results of searching for extremely high velocity outflows. To complete this endeavor, our team has developed a Python code in order to do this search systematically and have found 37 cases of quasars where the spectra present extremely high velocity outflows.

Introduction

Our solar system lies on an arm of the Milky Way, our spiral galaxy. Our Milky Way has a supermassive black hole in the center, just like most, if not all, galaxies. In some galaxies, the presence of gas around the supermassive black holes at their centers makes then an active galactic nuclei (AGN), which we can observe at large distances due to their large luminosity, which is the amount of light they emit. In the most massive elliptical galaxies, we find the black holes with the largest masses (around 50 billion times our Sun's mass!) and thus the most luminous AGN – quasars. When quasars were first observed they were given the name quasi-stellar objects because of their resemblance to stars; later this name was shortened to "quasars".

Quasars are some of the most luminous objects that reside in the core of the most massive galaxies, and quasars are so distant that they visually resemble stars. Quasars are some of the most distinct objects in our night sky [3]. Visually speaking, they are faint due to their distance from Earth. However, they are among the most luminous, and therefore energetic, objects in the night sky based on the amount of light they emit. Along those same lines, another unique property that distinguishes quasars from other stellar bodies is the fact that they radiate over the entire spectrum.

We are interested in the quasars' outflows; these are composed of matter that is being ejected away from the black hole environment. In these matter outflows there are ionized atoms (such as CIV, which is Carbon ionized three times) that are expelled out as they make their way into the black hole. These outflows are probably related to gas in the accretion disks,which are composed of matter that is orbiting and eventually falling into the black hole. Outflows may have an impact on star formation and galaxy evolution in their surroundings.

Our study aims to determine the number of quasars that present extremely high velocity outflows (EHVO – those outflows with speeds larger than 10% the speed of light [4]), and possible correlations between the presence of these outflows and other properties. To answer this scientific question, our group analyzed 6760 quasar spectra from the Sloan Digital Sky Survey Data Release 9 (SDSS DR9) [2], a public data set, searching for EHVO.

Data

We used two data sources for this study. SDSS DR9, an optical survey of objects in the night sky, and the Very Large Array Faint Images of the Radio Sky at Twenty-Centimeters (VLA FIRST), which is a survey that provides the radiometric data and covers mostly the same region of the sky as SDSS. Both data sets are public. We then were able to cross correlate the data from SDSS DR9 and FIRST by using the SDSS DR9 quasar catalog [5]. One problem we have encountered with using data from SDSS is that the data set can be biased since some objects are targeted for observation based on their radio properties [6].

We worked with the SDSS DR9, a public database containing 87 822 spectral targets. Of those quasars, 78 086 are new discoveries as of 2012. With the two cutoffs explained in further detail in the Methodology, we were able to reduce the number from 78 086 to 6760 quasars.

We then worked with our subsample of 6760 quasar spectra looking for outflows; we selected the brightest ones and the ones at the right distance from Earth. We have developed a Python code in order to do this search more efficiently. Outflows appear as absorption features, and to correctly identify absorption, we need to normalize the quasar spectra first. Our Python code fits a powerlaw to do so (as done by [4]). We subsequently carried out a visual inspection of the normalized data to verify that there was good normalization. In particular, we were interested in gas outflowing at the largest speeds (v > 30 000 km/s, 10% the speed of light!).

Methodology

Establishing CIV Absorption

The starting point was going through the plotted quasar spectra looking for signs of CIV absorption. For this to happen, we needed to establish two cutoffs. The first was making sure that the signal to noise ratio was high. With this the spectra are generally clearer and as a result we can have a more accurate absorption detection rate. And secondly, the redshift must also be 1.9 in order for the CIV and Lyα (Lyman-alpha, first emission line in the Lyman series of Hydrogen) to be included in the sections we are were observing. To solve for the redshift we used the equation,

$$z = \frac{\lambda_{\text{observed}} - \lambda_{\text{lab}}}{\lambda_{\text{lab}}}$$

Where $\lambda_{\text{observed}}$ is a value that we get from each individual quasar, and λ_{lab} is the literature wavelength value for CIV.

With these two cutoffs we reduced the number of quasar spectra from 87 822 to 6760.

Normalization of the Quasar's Continuum

With the 6760 quasar spectra, my research partner modified a spectrum normalization program originally written by a past collaborator [7] to fit a powerlaw to each quasar spectrum and normalize it. A separate spectrum analysis program was then used to flag any cases that had signs of broad absorption features, since

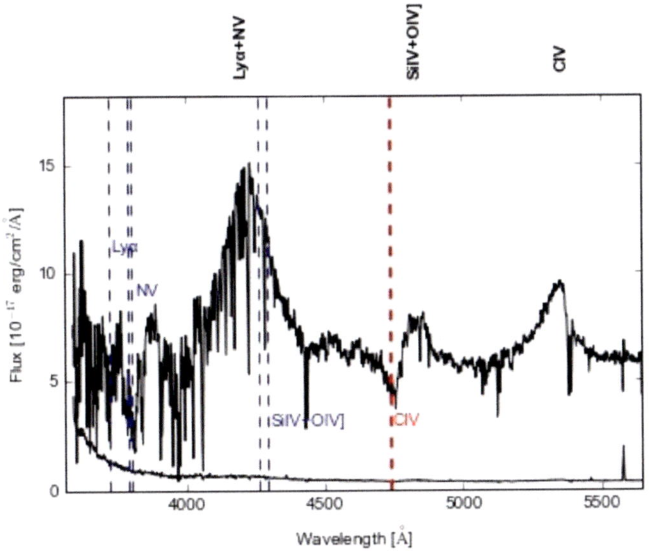

Figure 1. *spec-3826-55563-0860.dr9* Flux vs. Wavelength plot. Blue labeled lines represent possible ion absorption that is accompanying the CIV absorption, which is clearly indicated with red. The ion name is in black at the top of the plot and the labels are of the apparent emission lines.

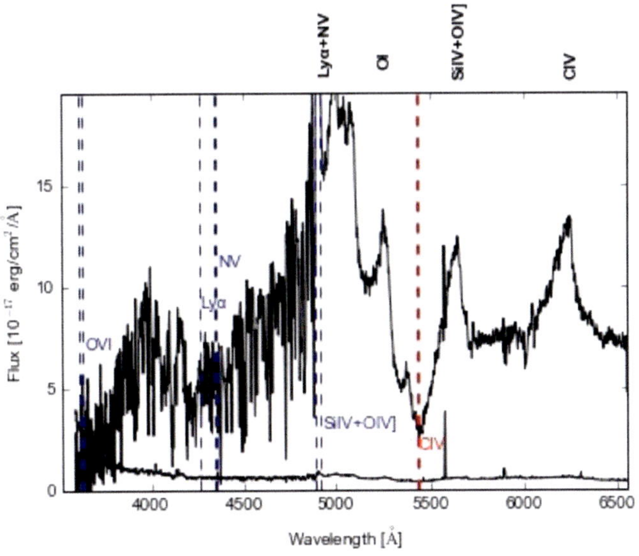

Figure 2. spec-4181-55685-0543.dr9 Flux vs. Wavelength plot. Reference **Fig. 1** caption. This is an example of a quasar with broader CIV absorption (larger width) than **Fig. 1**.

those are the cases relevant to this study. With the flagged normalized quasar spectra plots, we were able to visually inspect those spectra, searching for those that clearly had broad (widths > 1000 km/s) absorption of CIV. After reviewing the quasars, we found 37 cases that have possible CIV absorption. We are currently in a stage of reviewing the previously normalized spectra and preparing a program to manually normalize some quasar's spectra. We are keeping our figures very conservative and to ensure that, we are individually normalizing some questionable quasars to affirm that the CIV absorption really exists.

Preparing Our Data for Publication

Once we completed the visual inspection, we found 37 cases that have confirmed EHVO CIV absorption. Our team created a Python program to plot all the quasar spectra individually, which was tailored for each of the 37 cases that we found. The motive behind this was to ensure that we had a clear visualization of the CIV absorption and any possible accompanying absorption. This was very crucial step because: (1) we want these data to be public and accessible for others who are possibly conducting research on EHVO quasars and (2) if others can reference our database we might be able to discover more trends with this classification of quasars. Two examples can be found in **Figures 1** and **2**. The code had a function where we were able to input the wavelength value of CIV into the program and also align the rest of the ion emissions to their respective locations on the quasar spectra.

Conclusion

Once we completed the visual inspection, we found 37 cases where we have confirmed EHVO CIV absorption.

To make these findings public, our group will be launching a website in late 2018 where the data and plots of these EHVO quasars will be available. In that website the astronomy community will have access to: (1) the EHVO quasar information, (2) plots of all the cases found, (3) outflow information (e.g. width, depth), (4) presence of other ions in the outflow, and (5) links to all the SDSS information on these objects.

Once this database is available, our group and the entire astronomy community will be able to study possible correlations with other properties or look for trends among the EHVO quasar properties. For example, our team is studying whether the presence of EHVO outflows requires the presence of jets, which are radio emission components present in 10% of all quasars. Of the cases found to date, only 3 were found to be radio loud. Typically, 10% of all quasars are radio loud [1]. As of now this implies that radio loudness is not a prerequisite for EHVO. We will be presenting that work in another paper [8].

Future Work

We will continue the process of individually correcting the normalization on some of the quasar spectra in order to be able to upload the corrected plots onto our website. This website will include access to normalized plots as well as the original data. Once we have completed the website, we will expand to the most recent data release, SDSS DR14, in order to find more cases which will increase our statistical sample. This larger sample will allow us to investigate possible trends in black hole masses and quasars with EHVO.

Acknowledgements

We would like to acknowledge our advisor, Dr. Paola Rodriguez Hidalgo, as well as Abdul Moiz Khatri and Pat Hall.

References

1. Netzer, H. (2013). The Physics and Evolution of Galactic Nuclei, Cambridge University Press.
2. Paris, I. (2012) The Sloan Digital Sky Survey quasar catalog: ninth data release, Astronomy and Astrophysics.
3. Peterson, B.M. (2008) The Central Black Hole and Relationships with the Host Galaxy, New Astronomy Reviews, Volume 52, Issue 6, p. 240-252.Rodríguez Hidalgo, P., Hamann, F., & Hall, P.(2011) The extremely high velocity outflow in quasar PG0935+417, Monthly Notices of Royal Astronomical Society, Volume 411, Issue 247.
4. Shen, Y. (2011) A Catalog of Quasar Properties from Sloan Digital Sky Survey Data Release 7, The Astrophysical Journal.
5. Ross, N. (2012) The SDSS-III Baryon Oscillation Spectroscopic Survey: Quasar Target Selection for Data Release Nine, The Astrophysical Journal Supplement Series, Volume 199, Issue 3.
6. Khatri, A.M. (2015) Extremely High Velocity Outflows, Research Paper for CTA395, University of Toronto.
7. Haas, H., Quintero, C., Rodriguez Hidalgo, P. (2018) Searching for a Connection Between Radio Emission and UV/optical Absorption in Quasars, ideaFest Journal, Volume 2.

Mathematics

Teaching Apportionment

Charles M. Biles[1]*

[1]Professor Emeritus, Department of Mathematics, Humboldt State University, Arcata, CA
*author correspondence: nia31415@comcast.net

Introduction

An objective of several university courses is to present a variety of current-interest topics that utilize mathematical thinking. One such topic is apportionment, defined as the process of distributing a fixed number of indivisible resource units among competing groups according to some measurable group asset. A featured application is congressional apportionment: how many seats in the U.S. House of Representatives should each state get based on the decennial census and constitutional guidelines [3], [6], [7], [10]. Congressional apportionment has two different approaches: constituency and House size. The constituency approach starts with the question, how many people should a congressperson represent? The House size approach starts with the question, how many seats should there be in the House? The constituency approach was used for reapportionment of the House based on the census years 1790–1840 [1], [2], [4]. However, the constituency approach does not lead to a fixed resources distribution problem. Hence, most mathematics texts contort the colorful history of congressional apportionment based on the first six censuses by forcing it into the House size approach which does yield a fixed resources distribution problem. This results not only in errors in portraying the historical record but also in a missed opportunity to present a rather dazzling application of some really basic mathematical problems.

An Average Lesson

To set the mathematical props on the stage of congressional apportionment, a class lecture should be devoted to two basic tasks: averaging and rounding. Suppose that $0 \le a < b$. What is the average of a and b? American history of congressional apportionment supplies five answers [2]. Denote the average of a and b by ave(a, b). Then, ave(a, b) =

1.	$\max(a, b)$	maximum of a and b
2.	$\min(a, b)$	minimum of a and b
3.	$AM(a, b) = (a + b)/2$	arithmetic mean of a and b
4.	$HM(a, b) = 2ab/(a + b)$	harmonic mean of a and b
5.	$GM(a, b) = \sqrt{ab}$	geometric mean of a and b

Each of these averages can be applied to the problem of how to round a decimal. Suppose $q > 0$ with integer part n where $q - n > 0$. Denote the rounding of q by round(q). Then round(q) $\in \{n, n+1\}$ where round(q) = $n + 1$ if and only if:

1.	$q \ge \max(n, n+1)$	round down since this criterion is never satisfied
2.	$q \ge \min(n, n+1)$	round up since this criterion is always satisfied
3.	$q \ge AM(n, n+1)$	round normally
4.	$q \ge HM(n, n+1)$	harmonic mean rounding
5.	$q \ge GM(n, n+1)$	geometric mean rounding

The Basic Divisor Method

Let $U = \{S_1, S_2,..., S_N\}$ be a federal union of N states (N is a natural number, $N \ge 2$). Let $< p_1, p_2,..., p_N >$ denote the census; i.e., p_i is the population of state S_i. The congressional apportionment problem is to determine an apportionment vector $< a_1, a_2,..., a_N >$ where each a_i is a natural number. The census is necessary to follow the constitutional mandate that apportionment among the states be "according to their respective numbers" as enumerated by a decennial census.

A constituency approach to congressional apportionment naturally leads to the basic divisor method which applies a 3-step algorithm.

Step 1.	How many people should a congressperson represent? Answer: d
Step 2.	Calculate each state's **quotient**: $q_i = p_i/d$
Step 3.	Let $a_i = \max(1, \text{round}(q_i))$

Step 3 is formulated to satisfy the constitutional requirement that each state receive at least one seat in the House. Each apportionment act based on the censuses from 1790 through 1840 used this 3-step algorithm. The acts from 1790–1830 rounded the quotient by rounding down. Three alternatives were proposed during debates based on the 1830 census: John Quincy Adams, round up; James Dean, round up if and only if $p_i / (n_i + 1)$ is closer to d than p_i / n_i; Daniel Webster, round normally. Dean's proposal is mathematically equivalent to harmonic mean rounding while Webster's proposal is equivalent to arithmetic mean rounding [1], [2]. The apportionment act based on the 1830 census continued tradition by rounding down. The act based on the 1840 census rounded normally. Hence, by the time of the apportionment act based on the 1840 census there were four variations of the basic divisor method. These variations, each essentially concerned with how to round a decimal, are identified with a historical reference as follows.

Jefferson	round down
Adams	round up
Webster	round normally (use arithmetic mean rounding)
Dean	round using the harmonic mean criterion

The Quota Method

Note that the House size is merely the result of the basic divisor method; hence, a constituency approach to congressional apportionment does not lead to a fixed resource distribution problem. Thus the historic congressional apportionments based on the censuses 1790–1840 are not applications of apportionment as defined in modern texts. The first apportionment act to apply the fixed resource distribution definition was based on the census of 1850 which set the House size, h, at 233. After setting h Congress applied the quota method, a method based on the natural premise that if a state has x% of the population, then it should have x% of the seats in the House. The quota method utilizes a 4-step algorithm.

Step 1.	Determine the House size, h
Step 2.	Calculate each state's **quota**: $Q_i = h(p_i/p)$ where $p = \sum p_i$
Step 3.	Let L_i be the integer part of Q_i. Initialize $a_i = L_i$
Step 4.	Create a priority list to distribute the remaining $h - \sum L_i$ seats

The quota, Q_i, represents a state's "fair share" of h seats based on its share of the national population, p. Invariably Step 3 distributes most but not all of the seats and one is faced with the situation that $0 < h - \sum L_i < N$. The remaining $h - \sum L_i$ seats are distributed by means of a priority list. American history has offered the following options for this priority list [1], [2].

Hamilton	$Q_i - L_i$
Lowndes	p_i / L_i
Hill	$p_i / GM(L_i, L_i + 1)$

Hamilton's quota method is the only variation in American history ever applied to formulate an apportionment act based on the quota method.

The Modified Divisor Method

Congress abandoned the basic divisor method after the apportionment act based on the 1840 census primarily because the method suffered from rampant political gamesmanship. Congress abandoned the quota method after the discovery of deal-breaking paradoxes, especially the Alabama Paradox [1]–[4], [6]–[9]. The basic divisor method is based on the constituency approach to congressional apportionment while the quota method is based on the House size approach. Since these are the only two approaches to the congressional apportionment problem, Congress sought to blend the two methods in a way that would avoid their worst results. Accordingly, Congress adopted the modified divisor method which utilizes a 5-step algorithm.

Step 1.	Determine the House size, h
Step 2.	Initialize the divisor d with p/h (p is the national population)
Step 3.	Calculate each state's quotient: $q_i = p_i / d$
Step 4.	Let $a_i = \max(1, \text{round}(q_i))$
Step 5.	If $\Sigma a_i = h$, then done; else modify d and go to Step 3

The modified divisor method is merely the basic divisor method with a predetermined answer. Textbooks refer to the initial divisor calculated in Step 2 as the standard divisor [3], [6], [7], [10]. One calculates the standard divisor as a reasonable value to initiate the divisor algorithm; however, it usually does not produce the desired House size, h, in Step 5. Accordingly, this value for d must be adjusted (modified) in order to obtain the specified value for h.

Variations occur in Step 4 where one must choose a rounding technique. In addition to the four rounding techniques inherited from the basic divisor method, another was introduced during discussions based on the 1910 census. Edward Huntington advocated rounding based on the geometric mean, the same criterion Joseph Hill used to create a quota method priority list. Accordingly, this variation is called the Huntington-Hill method.

Many of today's mathematics writings refer to Jefferson's, Adams's, Dean's, Webster's, and Huntington-Hill's methods only in the context of a modified divisor method [3], [6], [7], [8], [10]. These adjectives only specify the rounding technique and can serve this purpose for both basic and modified divisor methods. It is noteworthy that current congressional apportionment law specifies the Huntington-Hill modified divisor method with $h = 435$ [4].

Priority Techniques

The modified divisor method accomplishes the goal of avoiding the worst problems of the basic divisor method and the quota method. However, the modified divisor method was presented applying an ad-hoc algorithm specific to a given House size. If one wants to compare the results with other House sizes, then one needs to rerun the algorithm for each size of interest. Accordingly, the Census Bureau developed a serial technique for distributing seats in the House [4]. First, each state is given one seat. This complies with the constitutional requirement that each state must have at least one seat. The Constitution further specifies that House seats are to be based on population. Today, giving one seat to each state distributes 50 seats. In a serial approach for further distribution, we ask, which state has priority for the 51st seat? 52nd seat? 53rd seat? Etc. In general, if a state has n seats, what is its priority for gaining an additional seat?

In response, let PN(n) be the priority number for a state to receive an (n+1)st seat given that the state has n seats. We define $\text{PN}(n) = p_i / \text{ave}(n, n+1)$. We then achieve each of the five modified divisor methods by setting ave(n, n+1) as follows [1], [2], [4].

Jefferson	$\max(n, n+1)$
Adams	$\min(n, n+1)$
Webster	$\text{AM}(n, n+1)$
Dean	$\text{HM}(n, n+1)$
Huntington-Hill	$\text{GM}(n, n+1)$

Today the Census Bureau calculates priority values for seats 51 through 440 using the Huntington-Hill method. Since current law specifies 435 seats, based on the 2010 census seat 434 went to California, seat 435 to Minnesota, and seat 436 would have gone to North Carolina [5].

The Classroom

The congressional apportionment problem is a magnificent problem to incorporate not only into liberal arts mathematics courses but also secondary education teacher training courses. A key point of this paper is that the American history of this problem acts as a driver and motivator for the mathematics. Accordingly, using the standard 50-minute class length as a model, it works well to devote five days to apportionment as follows.

Day 1.	An average lesson
Day 2.	The basic divisor era: 1790 – 1840
Day 3.	The quota method
Day 4.	The modified divisor method
Day 5.	Priority computation techniques

Open-source materials for these topics are available on the author's websites [11]. Day 1 establishes the skills needed for apportionment calculations. It also leaves the student with a "what's this stuff good for?" feeling that is satisfied in Days 2–5 where the five averaging and rounding mechanisms are applied to a real problem in American history. Day 2 focuses on the basic divisor method which establishes the platform for studying fixed-resources distribution problems. Congressional apportionment serves to motivate the evolution of mathematical thinking about apportionment rather than merely serving up examples.

Epilogue

The history of congressional apportionment serves well as background and motivation for a comprehensive treatment of apportionment in general. The congressional apportionment problem is easy to state but challenging to resolve. Resolution first requires a choice of approach: constituency or House size. A constituency approach naturally led to the basic divisor method. The House size approach first led to the quota method and then to the modified divisor method. These approaches produced the Jefferson, Adams, Dean, Webster, and Huntington-Hill divisor methods along with the Hamilton, Lowndes, and Hill quota methods. Many mathematics textbooks, however, treat apportionment solely as a fixed resources distribution problem, thereby ignoring the constituency approach resulting in errors in presenting the historical record.

Divisor methods introduced the problem of how to round a decimal and subsequently the challenge of how to create a priority list. At the foundation is the question of how to average two numbers. Although averaging two numbers and rounding a decimal may sound trivial at first, they lead to substantial situations demanding in-depth analysis making apportionment an ideal liberal arts topic. The depth of the subject is portrayed by the stunning Balinski-Young Impossibility Theorem: there are no perfect apportionment methods—any divisor method is subject to quota violations and any quota method is subject to paradoxes [1]. Accordingly, the Balinski-Young Theorem is to apportionment what Arrow's Theorem is to voting theory.

One may conclude a presentation of congressional apportionment with a view to the future since some change in current law is inevitable. Possible reform ideas include the Wyoming rule, the proposals of thirtythousand.org, and the proposal of Neubauer and Gartner [9], changing the House size, or simply replacing the Huntington-Hill criterion for rounding by Webster's [1], [2].

An alternative and engaging conclusion is to highlight the connection of congressional apportionment to the electoral system of selecting the President and Vice-President of the United States. A debate featuring The Electoral College vs. A Popular Vote provides a lively arena to connect voting theory and apportionment with aspects of journalism, politics, government, history, and mathematics.

Acknowledgements

Professors John Martin and Dan Munton at Santa Rosa Junior College, Professors Dale Oliver, Adam Falk, Tim Lauck, Kamila Larripa, and Holland Heese at Humboldt State University, Professors Robin Carter and Levi Gill at College of the Redwoods, Ms. Ashlee Buczek at Academy of the Redwoods, and Professor Shannon Guerrero at Northern Arizona

University for their collaboration and encouragements in this apportionment classroom project.

References

1. Balinski, Michel L. and Young, H. Peyton, Fair Representation, 2nd edition, Brookings Institute, 2001.
2. Biles, Charles M. The History of Congressional Apportionment, Humboldt State University Press, 2017.
3. COMAP, For All Practical Purposes, 10th edition, W. H. Freeman, 2015.
4. http://www.census.gov.
5. https://www.census.gov/population/apportionment/files/Priority%20Values%202010.pdf.
6. Johnson, David B. and Mowry, Thomas A., Mathematics: A Practical Odyssey, 8th edition, Cengage Learning, 2016.
7. Lippman, David, Math in Society, 2014, http://www.opentextbookstore.com/mathinsociety/.
8. Malkevitch, Joseph, Apportionment: Introduction, http://www.ams.org/samplings/feature-column/fcarc-apportion1.
9. Neubauer, Michael and Gartner, Margo, A Proposal for Apportioning the House, Political Science and Politics, 44 no. 1 (2011): 77-79.
10. Tannenbaum, Peter, Excursions in Modern Mathematics, 8th edition, Pearson, 2014.
11. www.nia977.wix.com/drbcap (personal website); https://digitalcommons.humboldt.edu/apportionment/ (HSU Digital Commons website)

NATURAL SCIENCES

Survey and Map Distribution of English Ivy (*Hedera helix* L.) at Patrick's Point State Park, California

John Cortenbach[1], Buddhika D. Madurapperuma[2*]

[1]Undergraduate Student, Departments of Forestry and Wildland Resources and Environmental Science and Management, Humboldt State University, Arcata, CA

[2]Lecturer/Research Associate, Departments of Forestry and Wildland Resources and Environmental Science and Management, Humboldt State University, Arcata, CA

[*]author correspondence: bdm280@humboldt.edu

Abstract

Patrick's Point State Park has experienced a tremendous growth of English ivy (*Hedera helix* L.) that is causing damage to the park resources. The goal of this study was to accurately map English ivy habitats and estimate coverage and colonization effects on tree trunks within the park. The results showed that the English ivy growth has decreased from 8.0 acres to 6.5 acres between 2015 and 2016, respectively, due to park restoration activity supported by volunteer organizations. The English ivy growth on tree trunks was significant for western hemlock and Sitka spruce. With the completion of this project, the park will now be able to locate English ivy spots that require the most attention and monitor their growth rate.

Introduction

English ivy (*Hedera helix* L.) is an evergreen woody climber that produces adventitious roots to climb on a wide range of natural and artificial substrates (Melzer *et al.*, 2011). English ivy was introduced to Patrick's Point State Park in Trinidad, California. The plant is native to Northern Europe and was brought by European settlers in the late 1890s as an ornamental plant before the park acquisition. English ivy was so important to them that they wanted to maintain it for future generations. Patrick's Point has a historical past from a group these settlers called "the Brooks" (Van Vleck, 1983). This information was documented by the Resource Agency California Department of Parks and Recreation (2016).

Understanding the historical landscape of Patrick's Point State Park is important to assess how English ivy was first introduced to the park by the European settlers. Michelle Forys, a local environmental scientist, shared a photograph of the settlers in front of the Ceremonial Rock in 1895 (**Plate 1**; Van Vleck, 1983). The evidence of settlers living on the land shows the possibly of introducing English ivy to a new ecosystem. In this study, the potential homestead of the European settlers was modeled using the spread of English ivy growth by comparing the collection of point-marked data and the National Land Cover Database (NLCD, 2011) by the U.S. Geological Survey (USGS).

The potential homestead of the European settlers is useful to understand how English ivy was introduced into the ecosystem and the distance from the settler's homestead. The results shown in **Fig. 1** illustrate that the trees growing in their beginning stage during the late 1890s show the location of where the English ivy was planted near the most suitable developed open space. Thus, the European settlers grazed,

Plate 1. A photograph showing the European settlers in front of the Ceremonial Rock in 1895 before Patrick's Point State Park was established in 1929–1931(Van Vleck, 1983).

building a home and introducing English ivy. We estimated the best suitable land for planting English ivy using NLCD to classify "developed, open space", which commonly included large-lot single-family housing units.

According to the literature, English ivy's ecological niche in its native range is comprised of floodplains (Schnitzler and Heuzé, 2006), urban forests (Hawthorne *et al.*, 2015; Copp, 2014), beech woods (Metcalfe, 2005), and woodlands (Rodwell, 1991). The species has become widely distributed at urban parks (Baskin, 2002; Dlugosch, 2005), fragmented and logged forests (Butaye, 2001; Matthews *et al.*, 2016), and residential to wilderness areas (Reichard, 2000) in the Pacific Northwest. The ecological factors, which have contributed to the alarming growth of English ivy, are moist nutrient-rich substrates (Schnitzler and Heuzé, 2006), anthropogenic disturbances (Ramsey, 2005), and light and temperature (Copp, 2014). English ivy is a superior competitor for natural resources, thus it is essential to control its growth in natural and urban environments in order to facilitate native plant's regeneration.

Currently, English ivy has widely spread into the park's natural ecosystems and the state park is now looking to control its growth within the park premises. The goal of the state park is to remove English ivy while preserving the future ecosystem's native growth.

The objective of this study is to utilize geospatial science to precisely map English ivy habitats within the state park in order to implement an effective management practice. This study is important because it will provide useful information on English ivy habitats within the park, where park managers and other conservation biologists can manage its growth for restoration habitats for native trees. From a historical point of view, it is even more important in understanding how an invasive plant, such as English ivy, cannot only affect our ecosystem, but how or why it was introduced within the area of Patrick's Point State Park. Therefore, it is vital to monitor the habitats of this invasive plant. This work is a continuing step towards management and the preservation of our natural and native vegetation. The enjoyment of future generations will be to ensure ongoing protection of our native species and ecosystems.

Methods

Study Area

Patrick's Point State Park is located on the northern coast of California in Humboldt County, about 20.6 miles north of Humboldt State University. The park is accessible by exiting Highway 101 and heading along the right side of Patrick's Point Drive. The area of the state park is 640 acres with its boundary just out into the Pacific Ocean (http://www.parks.ca.gov/parkindex). The state park is known for its natural enjoyment for campers and hikers, preserving the aesthetic view of the coastline.

Figure 1. The homestead of European Settlers and the location of planted English ivy in the late 1890s.

In-situ Data

A reconnaissance survey was done to assess how English ivy is distributed at Patrick's Point State Park in different habitats. Species occurrence in open habitats, road edges, and forest interior habitats were preliminarily observed. In addition, English ivy's growth along the trunk of some tree species such as western hemlock (*Tsuga heterophylla*), red alder (*Alnus rubra*), Douglas-fir (*Pseudotsuga menziesii*), and Sitka spruce (*Picea sitchensis*) was observed. Preliminary data on English ivy's acreage gathered from the park managers were utilized for this study to estimate the English ivy's growth at park locations. In this survey, we utilized two sources of data: (i) field data collected from Trimble Juno handheld Global Positioning System (GPS) and Impulse Laser Rangefinder and (ii) National Agriculture Imagery Program (NAIP) high-resolution images and associated GIS data sources.

A Trimble Juno GPS was used to mark and create polygons and points of any cited English ivy growth. In addition, an Impulse Laser Rangefinder was used to measure the growth height of English ivy on the tree trunk (**Fig. 2**).

English ivy growth within the park was estimated using the collected data between 2015 and 2016. The English ivy spatial data (i.e. "Hedra helix_2015" shapefile) was provided by Environmental Scientist Michelle Forys of the North Coast Redwood District. We used the shapefile to compare the growth rate with newly collected data in July 2016. The authors hiked to 2015-marked English ivy land cover to estimate any change within a year. New polygons were created to catalogue a reduction of English ivy

Figure 2. Measuring tree height using an Impulse Laser Rangefinder.

growth.

Remote Sensing Data

Data acquisition

Digital satellite data were downloaded from GIS web portals as shown in **Table 1** and then the data were imported to ArcMap©. The aerial photographs were used to digitize the feature classes such as streams, roads, and hiking trails.

Image processing

The image processing procedures were used to aid the interpretation of remote sensing images. The image processing involved digital enhancement (i.e. manipulating the contrast between objects) and spatial filtering (to detect the edges between features thereby defining boundaries) in order to improve the interpretation of these images. Image processing entails the geo-referencing of the satellite and aerial photographs. All images were spatially referenced to the projection of NAD 1983 UTM Zone 10 N. Ground control points (GCPs) were collected using the Trimble Juno GPS and the images were geo-referenced using GCPs. It was necessary to use a GPS to collect GCPs because of the lack of many distinct and permanent landmarks in the park.

Classification of images

Image classification was undertaken in order to separate the spectral data into different categories and was used with reference to the various vegetation types in the study area. The USGS Earth Explorer was used to collect NAIP imagery in 2014 of the park. NAIP imagery published on 3 December 2014 was used to estimate the Normalized Difference Vegetation Index (NDVI), which represents greenness of vegetation, to understand where English ivy tends to spread.

NDVI is calculated using the following mathematical formula:

$$NDVI = (NIR - Red) / (NIR + Red)$$

When sunlight strikes objects, certain wavelengths within this spectrum are absorbed (Red visible spectrum), and other wavelengths are reflected near infrared (NIR). The National Aeronautics and Space Ad-

Table 1. Data sources for mapping English ivy distribution at Patrick's Point State Park.

Dataset name	File format	File title	Source
Patrick's Point	Shapefile	CA State Parks	http://data.california.opendata.arcgis.com
2014-2015 Slow IVY Growth	Shapefile	Ivy Treatment.shp	California Department of Parks and Recreation
2014-2015 Faster IVY Growth	Shapefile	Ivy Treatment.shp	California Department of Parks and Recreation
v2016 IVY Growth	Shapefile	Ivy Treatment.shp	California Department of Parks and Recreation
Data Elevation Model	DEM	Grdn42w125_13	http://viewer.nationalmap.gov/basic/?basemap=b1&category=ned,nedsrc&title=Elevation%20View#productSearch
NAIP Imagery	MrSID	Humboldt County	https://earthexplorer.usgs.gov/

ministration (NASA) states, "pigment in plant leaves, chlorophyll, strongly absorbs visible light (from 0.4 to 0.7 μm) for use in photosynthesis. The cell structure of the leaves, on the other hand, strongly reflects near-infrared light (from 0.7 to 1.1 μm)" (Weier and Herring, 2000). NDVI is an indication of the condition of green vegetation with values typically ranging from -1 to +1, with values > 0.5 indicating dense vegetation and values < 0 indicating bare ground.

Boundary determination

The potential homestead boundaries of European settlers in Patrick's Point State Park were mapped using NLCD, a digital elevation model, and existing English ivy locations. NLCD represents land-use/cover classes, for example, classes 21–24 are categorized as "developed" areas. In ArcGIS© 10.2.2, the Raster Calculator was used to select cells (n = 21) that meet the criteria for "developed, open space". The formula ""land_use_data" ==21" was entered to Raster Calculator to see the areas. The elevation, slope, and land cover data were used to locate the human-settled lands, where the land cover equals the selected cells

Figure 3. Map showing the location of English ivy habitats in Patrick's Point State Park in Humboldt County, California.

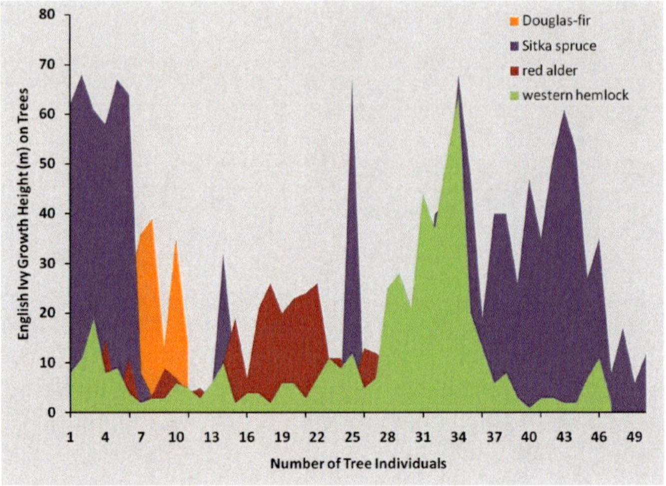

Figure 4. The height of English ivy growth on tree trunks vs. number of tree individuals (a total of 189 native trees were observed).

(n = 21) and the slope elevation. The European immigrants lived in the highlighted areas where the land was flat to build a home and graze landscapes (**Fig. 1**).

Data analysis

ArcGIS© 10.2.2 software was used for geo-spatial analysis and Trimble GPS PathfinderOffice® software was used to process collected *in-situ* data. The NDVI classification was done using ENVI® software. English ivy growth at Patrick's Point State Park from 2015 to 2016 is shown in **Fig. 3**.

Results & Discussion

English ivy abundance along the roads and the camping sites showed notable growth. English ivy growth at road edges could be caused by turbulent airflow around passing vehicles that may laterally move seeds towards the verge of the road (von der Lippe *et al.*, 2013). Compared to vehicle roads, hiking trails showed a lower abundance of English ivy growth. The total acreage of English ivy habitat in the park between 2015 and 2016 was 8.0 acres and 6.5 acres, respectively. The results show a decrease of significant growth of English ivy cover. The park controlled English ivy growth while implementing restoration activity supported by volunteers (HSU Natural Resource Club and local Humboldt County residents)

Figure 5. Marking data of Sitka spruce trees with English ivy growth along Point Drive Road.

Figure 6. English ivy growth at Abalone Campground.

and the California Conservation Corp.

In relation to a steady decline of English ivy habitat, however, native tree species such as western hemlock, red alder, Douglas-fir, and Sitka spruce are significantly invaded by English ivy growth. The selection of native trees estimated as the most impacted by English ivy growth are shown in **Fig. 4**. The impact levels were measured by how many trees in each species were invaded by English ivy and how far the ivy grew along the tree trunk. The estimated result shows that both Sitka spruce and western hemlock trees are significantly impacted by English ivy (**Fig. 4**). The reason why English ivy grows on these specific native trees is questionable, and their impact on trees' health needs to be investigated. However, previous studies on bark characteristics of host plants suggested that rough bark is more favorable for English ivy growth, protecting it from herbivore browsing (Hegarty and Caballe, 1991; Schnitzler and Heuzé, 2006). The bark characteristics may facilitate higher abundance of English ivy on tree trunks of Sitka spruce due to the prevalence of moisture and dust particles in flaky bark. Furthermore, Environmental Scientist Forys stated that allowing English ivy to grow around the bark of the tree creates competitive struggle between the English ivy and the tree for nutrients, water, and sunlight. By creating competition for the tree resources, English ivy can make a tree weaker (Schnitzler and Heuzé, 2006). This study is only based on the growth condition of English ivy, however, this can lead to future research on how physical and chemical characteristics of tree species enhance English ivy growth on the tree trunk. Sitka spruce is the most impacted tree, with the highest total count of English ivy growth (50 individuals) followed by western hemlock (47 individuals) and red alder (30 individuals). Douglas-fir was the least-recorded species (11 individuals) for English ivy growth (**Fig. 4**). **Table 2** shows the statistical growth rates and the ratio of the English ivy growth height based on the tree height measurement, while **Table 3** shows site-specific English ivy growth. English ivy growth along Point Drive Road and Abalone Campground is shown in **Figs. 5** and **6**.

The results in **Fig. 7** suggest that English ivy shows a faster growth rate in a healthier-vegetation environment with high visible sunlight. In a low vegetation-density environment, sunlight is more accessi-

Table 2. Estimates of the most significantly impacted native trees by English ivy growth based on measuring both the height of the tree and the height of English ivy shown in **Fig. 3**.

Plant	Ivy average height (m)	Ivy range (min–max) (m)	Tree average height (m)	Tree range (min–max) (m)
Red alder	11.17 ± 7.40	1 – 26	20.97 ± 10.53	7 – 42
Sitka spruce	25.78 ± 23.34	1 – 68	55.42 ± 22.77	7 – 105
Douglas-fir	26.64 ± 9.69	13 – 39	41.00 ± 3.07	35 – 45
Western hemlock	10.91 ± 12.34	1 – 51	44.79 ± 14.83	7 – 75

Table 3. The total count of marked trees invaded by English ivy growth within each campground.

English ivy cited by location	No. trees with English ivy growth
Agate Campground	83
Point Drive Road	79
Bishop Plane Group Picnic Area	16
Outside public campground boundary	6
Abalone Campground	5

ble to native growth. This can lead to further issues involving English ivy growth control. It will continue to spread from its current growth location until this plant gains access to sunlight. To determine the density of healthy vegetation on a patch of land, NDVI is useful to observe the ratio between absorption of visible light (Red) and reflection of near-infrared (NIR) sunlight by plants. As shown in Fig. 7, the NDVI ranged from 0 – 0.59, and the English ivy occurrence was noticeable at the forest fringe, open campground, and the

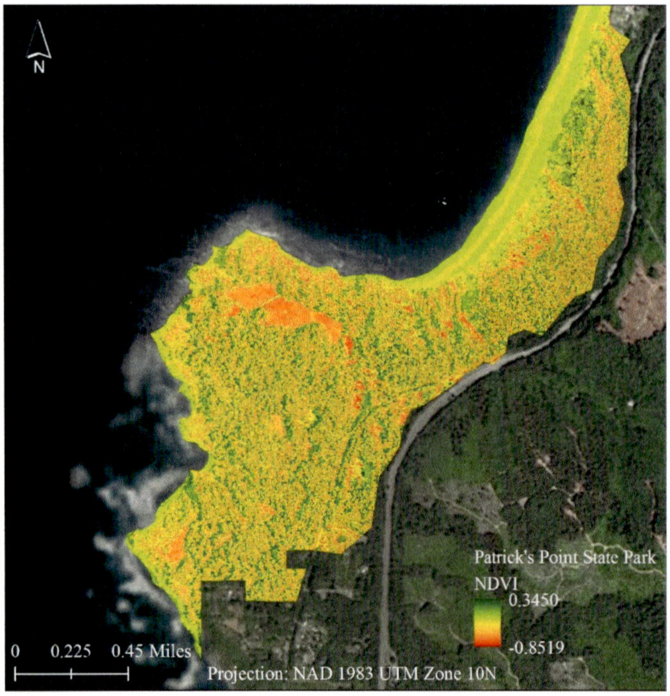

Figure 7. NDVI map of Patrick's Point State Park analyzing the density of plant growth of the state park using NAIP imagery 2014. This shows the ecological vegetation of possible access to sunlight. The greens indicate healthy closed vegetation, the reds indicate unhealthy, scarce vegetation or water, and the yellows indicate neutral vegetation. English ivy flourishes for sunlight visibility based on its location.

intact forest (Figs. 7 and 8). English ivy seeds are dispersed by birds from the tree trunk to the canopy gap area of intact forest (Soll, 2005). Seeds produced over 60 meters high on tree trunks enhance seed dispersal through seashore birds (e.g. Steller's Jay) that can potentially colonize in new habitats (Sulgrove, 2004; Swearingen and Diedrich, 2006). Therefore, controlling English ivy growth on the tree trunk before its flowering season is important in order to reduce the dispersal distance of seeds before they colonize in new habitats.

The invasive plant's growth rate based on the park management schedule of removing it once per month is shown in Fig. 8. The park announces restoration activities for volunteers and the support of California Conservation Corps throughout each season (winter, spring, and fall), except summer. The reason why summer is not added to their schedule is because most of the support from Humboldt State students and club members from the HSU Natural Resource Club are on vacation. This map allows the state park to manage their time and resources more sufficiently.

Markers illustrated in the legend correspond to the following:

White marker: 2014 – 2015 Slow Growth

- English ivy showed a minor increase of growth from 2014 and had not expanded significantly, resulting in less restoration activity within that certain area based on the measurements from shapefiles 2015-Hedera helix and 2016-Hedera helix.

Red marker: 2014 – 2015 Faster Growth

- English ivy land cover showed a significant increase in growth from 2014 to 2015 and had not decreased from constant restoration activity. Based on the measurements from shapefiles 2015-Hedera helix and 2016-Hedera helix.

Yellow marker: 2016

- Current English ivy restoration removal a month after the last data collection in

2016-Hedera helix shapefile. Last data collection of English ivy growth was on 17 October 2016.

The lack of management of an invasive plant caused a dramatic effect on the ecosystem. When non-native species are introduced into a new environment, their population may sometimes explode in numbers.

Figure 8. A restoration map illustrating the management of removing English ivy by hand from 2014 to 2016 (Forys, unpublished).

Conclusion & Future Work

Over the course of one year, Patrick's Point State Park is making significant progress in removing 0.5 acres of English ivy from the landscape. Now, sections of the native species are free to grow. This study can be used to further answer more research questions, such as how has this invasive plant continued to expand, and is there any competition? The growth of this invasive plant, which strangles native trees, must be managed before its blooming season, otherwise seeds will spread and begin a new cycle.

Acknowledgements

We would like thank to Michelle Forys, Environmentalist, North Coast Redwoods District, for sharing the species data within the park in previous years and George Pease, Stockroom Manager for providing the equipment for this research. We also thank Carolyn Delevich for making valuable suggestions and editing on the manuscript.

About the Authors

John Cortenbach is an undergraduate student in the Environmental Science Program at HSU. The paper is based on Cortenbach's capstone project. He is passionate about volunteering for social activities, hiking, and conducting GIS/remote sensing projects.The author is interested in managing English ivy populations through volunteering to remove the ivy, and Patrick's Point State Park needs more volunteers to take part in the conservation effort.

Dr. Buddhika Madurapperuma is a Lecturer/Research Associate in the Departments of Forestry and Wildland Resources and Environmental Science and Management. He is a major advisor for Cortenbach's capstone project. He teaches GIS, remote sensing, forest ecology, and dendrology classes at HSU. Dr. Madurapperuma conducts multidisciplinary research on remote sensing (i.e. hyperspectral remote sensing for invasive species detection and mapping), ecological studies on invasive species, and forest silviculture management.

References

Baskin, Y. 2002. The greening of horticulture: new codes of conduct aim to curb plant invasions. AIBS Bulletin, 52(6): 464–471.

Butaye, J., Jacquemyn, H., and Hermy, M. 2001. Differential colonization causing non-random forest plant community structure in a fragmented agricultural landscape. Ecography, 24: 369–380.

California Department of Parks and Recreation. Patrick's Point State Park. California Department of Parks and Recreation, 1 Jan. 2016. Accessed on 12 Dec. 2016.

Copp, S.R. 2014. Community level impacts associated with the invasion of English ivy (*Hedera* spp.) in Forest Park: a look at the impacts of ivy on community composition and soil moisture. Thesis Paper, Portland State University.

Dlugosch, K.M. 2005. Understory community changes associated with English ivy invasions in Seattle's urban parks. Northwest Science, 79(1): 53–60.

Hawthorne, T. L., Elmore, V., Strong, A., Bennett-Martin, P., Finnie, J., Parkman, J., and Reed, J. 2015. Mapping non-native invasive species and accessibility in an urban forest: a case study of participatory mapping and citizen science in Atlanta, Georgia. Applied Geography, 56: 187–198.

Hegarty, E., Caballe, G. 1991. Distribution and abundance of vines in forest communities. In: Putz, F.E., Mooney, H.A. (Eds.), The Biology of Vines. Cambridge University Press, Cambridge, pp. 313–337.

Matthews, E.R., Schmit, J.P., and Campbell, J.P. 2016. Climbing vines and forest edges affect tree growth and mortality in temperate forests of the US Mid-Atlantic States. Forest Ecology and Management, 374: 166–173.

Melzer, B., Seidel, R., Steinbrecher, T., and Speck, T. 2011.Structure, attachment properties, and ecological importance of the attachment system of English ivy (*Hedera helix*).Journal of Experimental Botany, 63: 191–201.

Metcalfe, D.J. 2005.*Hedera helix* L. Journal of Ecology, 93(3): 632–648.

Ramsey, T. 2005. Invasion of English ivy (*Hedera* spp., Araliaceae) into Pacific Northwest Forests. Biology, University of Washington, Seattle, WA: Dissertation.

Reichard, S. 2000. *Hedera helix* L. In C. C. Bossard, J. M. Randall, and M.C. Hoshovsky, editors.Invasive Plants of California's Wildlands. University of California Press, Berkeley, CA, pp. 212–216.

Rodwell, J.S., ed. 1991. British Plant Communities, Vol. 1. Woodlands and Scrub. Cambridge University Press, Cambridge, UK.

Schnitzler, A. and Heuzé, P. 2006. Ivy (*Hedera helix* L.) dynamics in riverine forests: effects of river regulation and forest disturbance. Forest Ecology and Management, 236(1): 12–17.

Soll, Jonathan. 2005. Controlling English ivy (*Hedera helix*) in the Pacific Northwest, [Online]. In: Control methods--Invasive plant management. In: GIST (Global Invasive Species Team). Arlington, VA: The Nature Conservancy (Producer). Available: http://www.invasive.org/gist/moredocs/hedhel02.pdf [2018, April 2].

Sulgrove, S.M. 2004. Is Ivy Invasive? What known about Ivy. Ivy Journal, The American Ivy Society, 30: 16–17.

Swearingen, J.M.,Diedrich, S. 2006. Fact sheet: English ivy--*Hedera helix* L., [Online]. In: Weeds gone wild: Alien plant invaders of natural areas--Fact sheets. Plant Conservation Alliance's Alien Plant Working Group (Producer).

Van Vleck, 1983. Patrick's Point State Park preliminary General Plan. State of California - The Resources Agency, Department of Parks and Recreation, 1–54.

von der Lippe, M., Bullock, J.M., Kowarik, I., Knopp, T., Wichmann, M. 2013. Correction: human-mediated dispersal of seeds by the airflow of vehicles. PLoS ONE 8(8): 10.1371/annotation/50f98e02-214c-4314-b053-ca272101ead1. https://doi.org/10.1371/annotation/50f98e02-214c-4314-b053-ca272101ead1

Weier, J. and Herring, D. 2000.Measuring vegetation (NDVI & EVI).NASA Earth Observatory, Washington DC.

Developing a Coastal GIS Model of Sri Lanka to Pinpoint Areas at Risk of Tsunamis

John E. Dellysse[1], Buddhika D. Madurapperuma[2*]

[1]Undergraduate Student, Departments of Forestry and Wildland Resources and Environmental Science and Management, Humboldt State University, Arcata, CA
[2]Lecturer/Research Associate, Departments of Forestry and Wildland Resources and Environmental Science and Management, Humboldt State University, Arcata, CA
[*]author correspondence: bdm280@humboldt.edu

Abstract

This study examined land-use changes along the south-eastern coast of Sri Lanka before and after the 2004 tsunami to spot areas vulnerable to flooding and severe weather events like tsunamis. On December 26th, 2004, over 30,000 people lost their lives after an offshore earthquake caused a tsunami in Sri Lanka, which resulted in waves as high as 30 meters. A time-series vegetation change: (i) immediately after the tsunami between 2004 and 2005, (ii) pre-tsunami & long-term between 2004 and 2016, and (iii) post-tsunami & long-term between 2005 and 2016, were mapped using Landsat TM images. The resulting series of change detection models were then utilized to create a series of maps displaying considerable disturbance of vegetation patterns and agricultural activity along coastal and inland regions.

Introduction

This study examined the land-use changes along the south-eastern coast of Sri Lanka before and after the 2004 tsunami to spot areas vulnerable to flooding and severe weather events like tsunamis. On December 26th, 2004, 230,000 people, spanning 14 countries, lost their lives due to an offshore earthquake that led to a tsunami with waves reported as high as 30 meters (United Nations, 2007). The events of 2004 were powered by an unforeseen seismic event measuring 9.1–9.3 magnitude along the Indian-Burma boundary off the coast of Sumatra, Indonesia (UNESCO, 2009). When analyzing the effects of this incident, areas vulnerable to flooding and tsunamis can be identified due to their lower elevations relative to sea level, and topography. These coastal communities can become even more vulnerable due to vegetation loss and the introduction of invasive species, which can destabilize the coastal soil composition.

In the coming decades, these communities will be the primary indicator of ecological, agricultural, and sociological devastation. As global warming continues to fuel increased climate changes, the effects will manifest as more frequent and more extreme localized disasters. The resulting loss of more than 30,000 people in Sri Lanka (Miura *et al.*, 2006) was a key factor that motivated the Sri Lankan government to realize their limitations in addressing major climatic events and the severe impact this had on Sri Lanka. In 2005, efforts were made to mitigate future events from further devastating at-risk coastal communities along the southern coastal areas, ranging from Colombo to Trincomalee. Since then, the government of Sri Lanka has been working with the World Bank's

Disaster Risk Management Agency (DRM) to develop mitigation measures, such as government planning and regulation of coastal agriculture, removal of invasive species, and replanting of native species along Sri Lanka's southern coastline. The reduction techniques developed in Sri Lanka could easily be implemented to generate future models, mitigation measures, and recommendations for subsequent projects along the coastal communities of Humboldt County or anywhere along low-lying coastal regions around the globe.

Almost half of Sri Lanka's southern coastline was affected by sea water inundation and related debris contamination post-tsunami (Illangasekare, 2006). Much of the coastal and inland ecosystems were debilitated by salt water and a variety of other contaminants, which resulted in the 2005 Sri Lanka Disaster Management Act No. 13 and led to the establishment of the Disaster Management Centre (DMC) for risk management. The roads and infrastructure were already suffering from war damage and neglect before the tsunami and were left devastated. Because of this, it has taken over seven years and 530 million pounds (674 million U.S. dollars) to enact repairs to a fraction of the roads, railway tracks, and bridges that were damaged (New Civil Engineer, 2005).

A rapid assessment of coastal habitats' rehabilitation in the southern coastal Hambantota District was conducted 20 months following the disaster (Bambaradeniya *et al.*, 2006). Samples in different land uses and land cover classes such as mangrove, sea-shore, and coastal shrublands were taken, revealing good regeneration potential for lost coastal vegetation (**Fig. 1**). As coastal zone demarcation for vulnerabilities, an approximate two kilometer inland buffer (boundary from coastline to inland) was considered the threshold for what defines a coastal vegetation area in our model (Government of Sri Lanka, 1981; Kaplan *et al.*, 2009) and thus an appropriate geographic buffer for this study. The vegetation studies and the associated buffer zones were then used as reference for different vegetation classifications. This two kilometer buffer for coastal biomass estimation was used to assess the vegetation's vulnerability to act as a protective shield against flooding and to stabilize coastal soils (i.e. mangrove vegetation, shelterbelts, windbreaks) and the potential impacts of a future tsunami. The vegetation within the two kilometer buffer zone is then defined as a biological shield (bioshield) and is characterized by vegetation composed of trees and shrubs grown along the coasts to protect coastal areas (Selvam *et al.*, 2005).

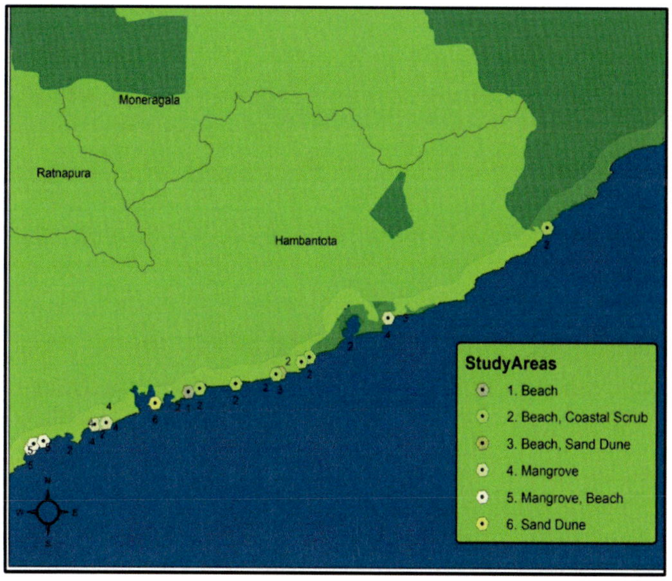

Figure 1. Post-tsunami coastal vegetation regeneration sites in Hambantota District. Map prepared using the ground control points surveyed by the IUCN (Source: Bambaradeniya *et al.*, 2006).

While coastal habitats are especially vulnerable to climatic change and tsunamis, 59% of the Sri Lankan population still lives in coastal districts with maritime boundaries (Nayanananda, 2007). Coastal vegetation belts, such as mangroves, have been shown to provide a natural buffer by absorbing the tsunami wave action and currents, thereby reducing human death from tsunamis (Kathiresan and Rajendran, 2005). Additionally, studies have measured the importance of sustainable management practices and the impact on tropical ecosystems such as mangrove forests, coral reefs, and seagrass beds (Dahdouh-Guebas, 2002).

The identification and assessment of stable coastal vegetation, or bioshield, along the coastline of Sri Lanka is useful for forecasting the specific areas at risk of large intermittent wave actions and the likelihood of risk to people and property, meaning it requires implementation of the best coastal management plans prior to coastal habitat rehabilitation. As a result, the vulnerability of these coastal areas to natural disasters is directly related to the risk management techniques of the Sri Lankan government (Jayawardane, 2006).

These coastal areas will continue to be vulnerable to sea level rise (SLR) along lands adjacent to wetlands and lowlands under the scenarios of 0.3 m and 1.0 m in coming decades (Weerakkody, 1996). According to UNDP (2007) SLR projections, the maximum SLR is estimated to be 0.59 m inundation. According to Madurapperuma *et al.* (2017), erosion at Oluvil beach has severely degraded coastal habitats and has led to a significant land mass inundation of 0.5 m and 2.0 m SLR. Predicted SLR in Sri Lanka will be 0.51 m in the next 25 years, 0.66 m in 50 years, and 0.96 m in 100 years (UNDP, 2007). The resulting sea water intrusion to wetlands and lowland paddy lands along these areas will likely destroy the ecosystem through inundation of saline water.

Since 2005 the government of Sri Lanka and the DRM have spent millions in the recovery effort by rebuilding coastal infrastructure and enforcing a 100 m buffer zone in the west and south and a 200 m buffer zone in the eastern and northern coastlines of Sri Lanka, restricting construction (Ratnasooriya, 2007). These efforts require an updated analysis of the coastal bioshield health to assess the effectiveness of the effort by the Sri Lankan government and the World Bank's DRM since 2005 (Jayawardane, 2006). A comprehensive change detection study needs to be implemented to assess mitigation efforts and measure the health and vegetation mass of the coastal bioshield.

The mangrove forests and coastal biomass need to be measured to better assess the effectiveness of the repair efforts as well as the return of biomass due to natural processes post-recovery efforts in 2005 along the south-western coastal regions from Colombo to the north-eastern coastal regions of Trincomalee. Additionally, a GIS-based assessment should be part of future studies as well, analyzing what coastal areas have not recovered and what coastal areas are still at risk of climate change-related wave action, saltwater contamination, and other occurring natural disasters.

This study focused on the current biomass of vegetation along the coastal region in and around Ampara Region, Sri Lanka using change detection within ENVI® image analysis software to develop a preliminary model. This model will later be expanded with *in-situ* data to update the DMC database on post-tsunami mitigation measures in Sri Lanka. Subsequent studies will develop a comparative analysis and model of tsunami mitigation techniques and coastal recommendations for the Northern California Humboldt coastline.

Methods

For this project a vegetation, or land-use, pre- and post-tsunami model for Ampara, Sri Lanka was developed (Fig. 2). The first step was to retrieve data

Figure 2. Locator map for Ampara (2017) and Hambantota (2006) IUCN study areas (Bambaradeniya *et al.*, 2006).

from Earth Explorer through the GLOVIS data visualizer for Landsat 5 TM and Landsat 8 with the dates: 12-2004, 04-2005, and 08-2016.

As depicted in the flow chart in Fig. 3, radiance to reflectance calculations were performed prior to analysis using the Band Math function in ENVI®. The brightness values, or radiance, of pixel values of the raw Landsat images were converted to reflectance using the calibrated reflectance values that can be found in the metadata of the Landsat images. The reflectance values were useful to delineate vegetation and moisture values of Landsat images and widely used to derive remotely-sensed indices. For example, vegetation health can be detected using the Normalized Differential Vegetation Indices (NDVI) (Madurapperuma *et al.*, 2017). When sunlight strikes vegetation, certain wavelengths of this spectrum are absorbed and other wavelengths are reflected. For instance, pigment in plant leaves, such as chlorophyll, strongly absorbs visible light (from 0.4 to 0.7 μm) for use in photosynthesis. The cell structure of the leaves, on the other hand, strongly reflects near-infrared (NIR) light (from 0.7 to 1.1 μm) (Weier and Herring, 2000). Near-infrared light refers to the main infrared component of the solar radiation (from 0.7 to 1.1 μm) reflected from the Earth's surface. The next step was to perform an NDVI analysis model within ENVI® to analyze biomass and health indicators for temporal elements (2004–2005, 2005–2016, and 2004–2016).

The bands used for the analysis were 3 (red band) and 4 (NIR band) for Landsat 5 and bands 4 and 5 (red and NIR band, respectively) for Landsat 8 imagery, with all results projected in WGS 1984 UTM zone 44N. The final comparisons performed were from pre-tsunami December 2004 to April 2005, from post-tsunami April 2005 to September 2016, and from December 2004 to September 2016.

Landsat 5 band description:

Band 3 - red 0.63 – 0.69 μm; Discriminates vegetation slopes

Band 4 - NIR 0.77 – 0.90 μm; Emphasizes biomass content and shorelines

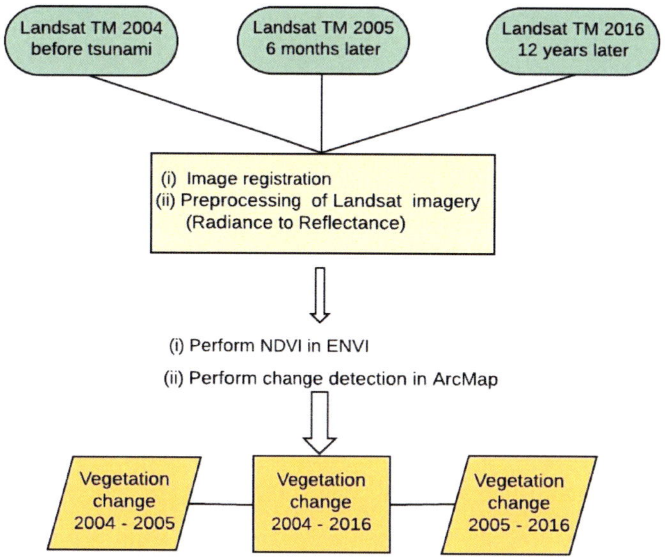

Figure 3. A change detection workflow for studying tsunami impact sites in Ampara, Sri Lanka.

Landsat 8 band description:

Band 4 - red 0.64 – 0.67 μm; Discriminates vegetation slopes

Band 5 - NIR 0.88 – 0.85 μm; Emphasizes biomass content and shorelines

The next step was exporting the images and data into ArcMap© for the development of all three maps. After importing into ArcMap©10.4 the NDVI raster images were then subset in ArcMap© using a two kilometer buffer to define functional coastline parameters (Government of Sri Lanka, 1981; Kaplan *et al.*, 2009). The resulting clipped NDVI images then underwent a pixel conversion from floating point to integer and were then reclassified by pixel values. The change detection process was then performed, resulting in three classifications:

a. All negative values or decreases below zero were classified as "-1"

b. All positive values or increases were classified as "1"

c. Any unchanged or zero values were classified as "0"

We then summarized the vegetation change and performed a spatial join of the features in terms of land-use classes in the Ampara study area to quantify the tsunami impact vulnerabilities for the best land-use practices in the future. The land uses within the two kilometer buffer were extracted and then vegetation change within each land use was estimated using the tabulate area function in ArcMap©.

Comparative graphs displaying increases or decreases in vegetation were then made to see how each land-use change was impacted by the tsunami. These values were then converted from raster to polygon layers to estimate area by cell size and pixel count. By comparing the difference in spectrally identified vegetation classes, a significant change in vegetation was identified post-tsunami in April 2005. The next comparison was for temporal elements of 2005–2016 which was done to develop a baseline model of what repair measures have been made since the 2004 tsunami. A change detection workflow chart between 2004–2005, 2005–2016, and 2004–2016 is given in **Fig. 3**.

Results & Discussion

After developing multiple change detection models spanning 12 years (**Figs. 4–6**), the resulting analysis shows a considerable amount of damage due to the 2004 tsunami, evidenced by vegetation loss due to saltwater contamination (Ratnasooriya, 2007). The NDVI is applied to detect areas of vegetation cover increase or decrease, which is given by positive and negative values and symbolized by green and red colors, respectively. The NDVI is useful to detect changes in vegetation over time. In this study, however, we used NDVI to detect vegetation health corresponding to high NDVI values in a particular time step and also

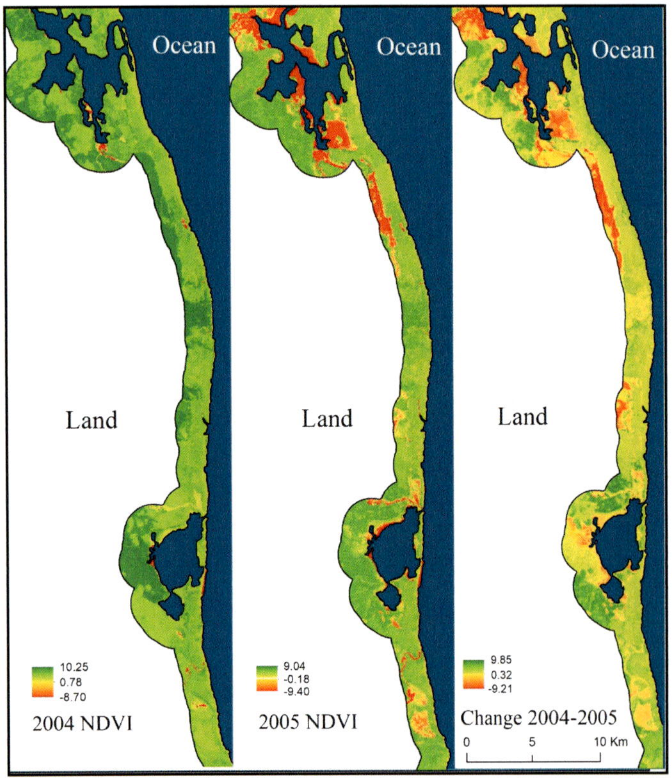

Figure 4. 2004 and 2005 NDVI images of Ampara pre- and post-tsunami with NDVI values (prior to change analysis) and 2004–2005 image with completed (short-term) change analysis. The 2004 and 2005 NDVI maps show vegetation health, where greens are healthy vegetation, reds are poor vegetation, and yellows are neutral vegetation health. In the change detection map, the greens are positive changes and the reds are negative changes, with yellow denoting few to no changes.

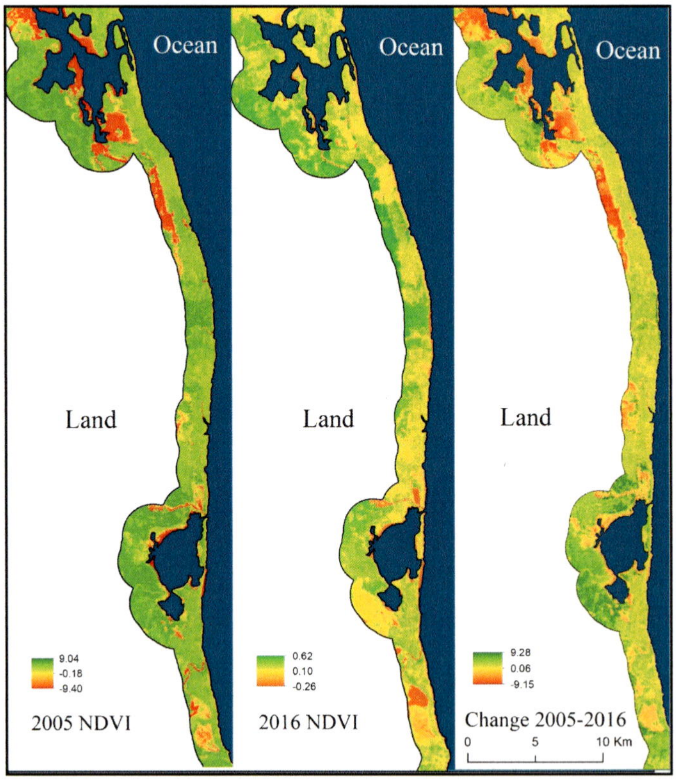

Figure 5. 2005 and 2016 NDVI images of Ampara over 11 years with NDVI values (prior to change analysis) and 2005–2016 image with completed (long-term) change analysis. The 2005 and 2016 NDVI maps show vegetation health, where greens are healthy vegetation, reds are poor vegetation, and yellows are neutral vegetation health. In the change detection map, the greens are positive changes and the reds are negative changes, with yellow denoting few to no changes.

detect vegetation changes between two time steps. As **Figs. 4–6** depict, some green areas in 2004 transformed to red in 2005 due to vegetation damage, or vegetation susceptible to tsunami wave impacts. A red area changing to green represents vegetation growth, or rehabilitation of coastal habitats. The change detection between 2004–2005 and 2004–2016 showed a significant decrease in vegetation mass post-tsunami (a more red color) compared to 2005–2016 (**Figs. 4–6**). This procedure was verified with a land-use database from the DATA.GOV (2015) and Google Earth for ground-truthing. The change detection analysis appeared to show cultivated areas were relocated further inland.

The coastal changes, both natural and cultivated, since December 26th, 2004 have resulted in a redistribution of natural vegetation and agricultural lands.

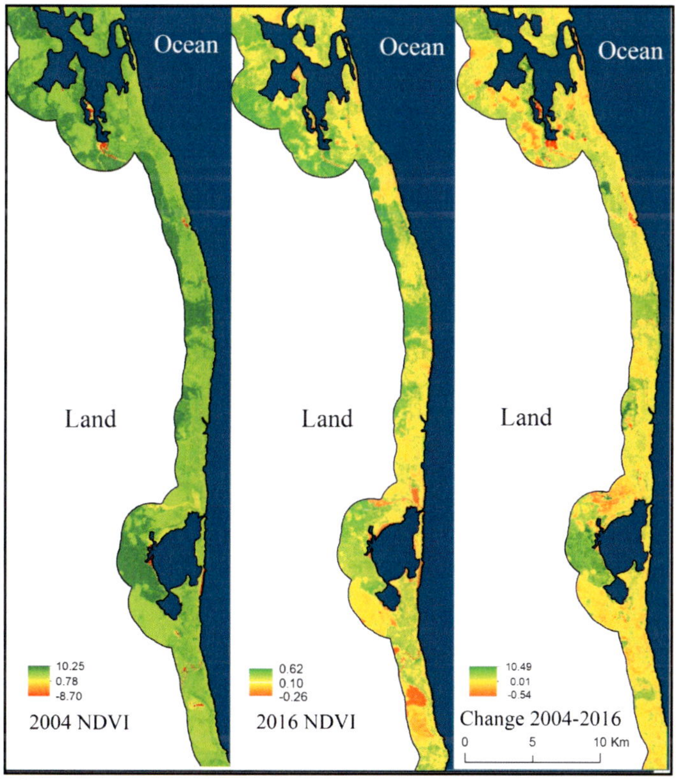

Figure 6. 2004 and 2016 NDVI images of Ampara over 12 years with NDVI values (prior to change analysis) and 2004–2016 image with completed (long-term) change analysis. The 2004 and 2016 NDVI maps show vegetation health, where greens are healthy vegetation, reds are poor vegetation, and yellows are neutral vegetation health. In the change detection map, the greens are positive changes and the reds are negative changes, with yellow denoting few to no changes.

The changes include the categories of tree cover (broad-leaved and deciduous) and closed vegetation. Much of these vegetation and cultivated areas appear to have relocated inland, likely due to the resulting soil contamination from seawater inundation (Mattsson *et al.*, 2009). This has left a lasting effect along the vegetation bioshields for tsunami mitigation from the coastal region of Ampara and immediately north to Batticaloa (Tanaka, 2009).

We summarized NDVI vegetation change in relation to land-use classes over time to see how each land-use class was impacted by the tsunami and how it recovered long-term with sustainable coastal management practices (**Fig. 7** and **Table S1**). The land-use change in relation to percent vegetation increased (i.e. natural and tree plantation for coastal habitat restoration) and/or vegetation decreased (i.e. tsunami and anthropogenic degradation, and sea water inundation) was then graphically presented by two bars for four land-use categories in **Fig. 7**. The raw figures on extents of per square acre land uses are given in **Table S1**. The land-use and land-cover changes derived from remotely sensed images were summarized into seven classes: dense forest, open forest, garden, other land, paddy, coconut, and undefined. There were no noticeable changes in open forest, coconut, and undefined (**Table S1**).

Of the land-use classes, natural vegetation, dense forests, and open forests have high resistance to tsunami wave actions, and thus, measured a lesser decrease in vegetation than anthropogenic land uses (i.e. garden, paddy, and other lands except coconut plantation) (**Table S1**). Therefore, implementing forests and trees as protective shielding of the coastline from tsunamis is effective (Fritz *et al.*, 2006). Mattsson *et al.* (2008) reported that natural forests and coconut plantations have contributed to high carbon sequestration in coastal ecosystems. In addition, extensive root mats of coconut trees offer protection from scouring and erosion (Forbes and Broadhead, 2007). Therefore, clear cutting coconut trees for harbor construction resulted in coastal erosion at Oluvil beach (Madurapperuma *et al.*, 2017).

According to the previous studies, buildings within one kilometer of the coastline in the eastern region have been severely damaged due to tsunami wave actions (Miura *et al.*, 2006). The maximum tsunami heights of the east coast were reported to be 3–7 m (Wijetunge, 2006). Therefore, transforming human-influenced land uses, such as paddy and home gardens, to natural coastal vegetation is beneficial to mitigating tsunami and cyclone effects. Based on these findings, we suggest that the government give incentives to landowners to convert their lands to more tsunami-resilient land uses.

Conclusion & Recommendations

After developing multiple change detection models spanning 12 years, the resulting analysis shows a considerable amount of devastation due to the wave action that inundated coastal communities from the 2004 tsunami. The changes and repairs since December 26th, 2004 resulted in the apparent relocation of agricultural lands, and the natural redistribution of vegetation further inland, which is likely due to the resulting soil contamination from seawater inundation (Mattsson *et al.*, 2009). This has left a lasting effect along the vegetation bioshields for tsunami mitigation (Tanaka, 2009), from the coastal region of Ampara and immediately north to Batticaloa. Additional research should be done using higher resolution imagery than the 30 m resolution provided by Landsat data. High resolution spatio-temporal data sets recorded from sequential aerial photography (i.e. unmanned aerial vehicle) would be essential for accurately mapping vegetation dynamics in Sri Lanka (Dahdouh-Guebas, 2002; Dahdouh-Guebas *et al.*, 2000). These additional data sets could then be utilized to identify which specific plant species were most and least affected by salt water inundation. To continue this process, high resolution images (i.e. drone images) could also be used to develop a more accurate digital elevation model to further identify lower elevation areas at risk of flooding. This could then be used to create SLR prediction models to identify specific waterways and corridors most vulnerable to SLR and inland salt water inundation. These

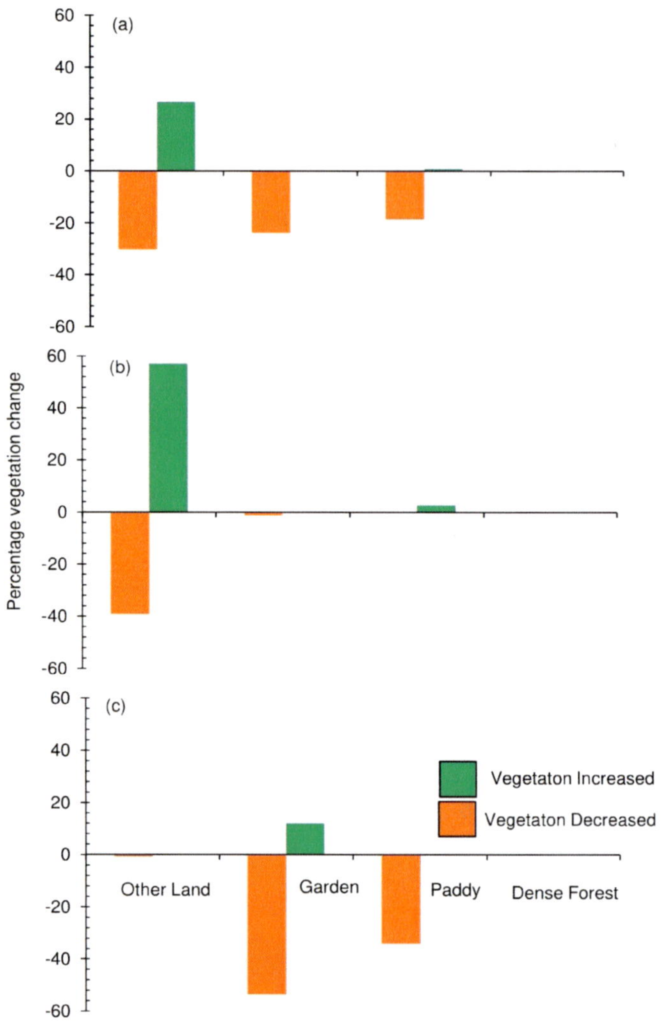

Figure 7. Vegetation biomass change in relation to land-use category in Ampara (a) 2004 –2005, (b) 2005–2016, and (c) 2004–2016.

improvements to the model would provide a highly accurate measurement of vegetation bioshields for the mitigation of tsunami wave action. The result would be a predictive model of inland coastal inundation patterns. These data could be developed further with existing SLR data with a statistical model measuring inland flooding patterns. The final result would be a study to repair and promote growth within at-risk coastal communities of Sri Lanka.

Acknowledgments

We would like to acknowledge Carolyn Delevich for making valuable suggestions and editing on the manuscript.

About the Authors

J.E. Dellysse graduated from HSU with a B.S. in Environmental Science in December 2017. He conducted this research for the intermediate remote sensing class project and further developed through his Directed Study advised by Dr. Madurapperuma. J.E. Dellysse visited Sri Lanka to collect the *in-situ* data to validate his change detection results over the summer of 2017.

Dr. Buddhika Madurapperuma is a Lecturer/Research Associate in the Departments of Forestry and Wildland Resources and Environmental Science and Management. He is a major advisor for J.E. Dellysse's Directed Study. He teaches GIS, remote sensing, forest ecology, and dendrology classes at HSU. Dr. Madurapperuma conducts multidisciplinary research on remote sensing (i.e. kite aerial photography and unmanned aerial vehicle), ecological studies on invasive species, and forest silviculture management.

References

Bambaradeniya, C., Perera, S., Samarawickrema, P. 2006. A rapid assessment of post-tsunami environmental dynamics in relation to coastal zone rehabilitation and development activities in the Hambantota District of Southern Sri Lanka. IUCN Occasional paper - No. 10.

DATA.GOV (2015). Integrated: Geospatial Toolkit GIS data for Sri Lanka from NREL. Retrieved from http://en.openei.org/doe-opendata/dataset/5eb97f54-f1cb-452b-828c-7ff318f280f1/resource/89d4faa0-e4ee-4a7c-8374 d6c-073955b7c/download/srilgstdata190.zip

Dahdouh-Guebas, F. 2002. The use of remote sensing and GIS in the sustainable management of tropical coastal ecosystems. Environment, development and sustainability, 4(2): 93-112.

Dahdouh-Guebas, F., Verheyden, A., De Genst, W., Hettiarachchi, S., and Koedam, N. 2000. Four decade vegetation dynamics in Sri Lankan mangroves as detected from sequential aerial photography: a case study in Galle. Bulletin of Marine Science, 67(2): 741-759.

Forbes, K., Broadhead, J. 2007. The role of coastal forests in the mitigation of tsunami impacts. Food and Agriculture of the United Nations, Regional Office for Asia and the Pacific Bankok.

Fritz, H.M., Borrero, J.C., Synolakis, C.E., Yoo, J. 2006. 2004 Indian Ocean tsunami flow velocity measurements from survivor videos. Geophysical Research Letters, 33: L24605, doi:10.1029/2006GL026784.

Government of Sri Lanka. Coast Conservation Act No. 57 of 1981.

Illangasekare, T., Tyler, S.W., Clement, T.P., Villholth, K.G., Perera, A.P.G.R.L., Obeysekera, J., Gunatilaka, A., Panabokke, C.R., Hyndman, D.W., Cunningham, K.J. and Kaluarachchi, J.J. 2006. Impacts of the 2004 tsunami on groundwater resources in Sri Lanka. Water Resources Research, 42(5): W05201, doi:10.1029/2006WR004876.

Jayawardane, A.K.W. 2006, March. Disaster mitigation initiatives in Sri Lanka. In Proceedings of International Symposium on Management Systems for Disaster Prevention. 09-11.

Kaplan, M., Renaud, F. G., & Lüchters, G. 2009. Vulnerability assessment and protective effects of coastal vegetation during the 2004 tsunami in Sri Lanka. Natural Hazards and Earth System Sciences, 9(4): 1479.

Kathiresan, K., Rajendran. N. 2005. Coastal mangrove forests mitigated tsunami. Estuarine, Coastal and Shelf Science 65(3): 601–606.

Madurapperuma, B.D., Dellysse, J.E., Zahir, I.L.M. and Aathamlebbe, I. 2017. Mapping shoreline vulnerabilities using kite aerial photographs at OluvilHarbour in Ampara.7th International Conference of South Eastern University of Sri Lanka, 193-204.http://ir.lib.seu.ac.lk/handle/123456789/3010

Mattsson, E., Ostwald, M., Nissanka, S.P., Holmer, B. and Palm, M., 2009. Recovery and protection of coastal ecosystems after tsunami event and potential for participatory forestry CDM–Examples from Sri Lanka. Ocean & Coastal Management, 52(1): 1-9.

Miura, H., Wijeyewickrema, Inoue. 2006. Evaluation of tsunami damage in the eastern part of Sri Lanka due to the 2004 Sumatra earthquake using remote sensing technique. Proc. 8th National Conference on earthquake Engineering, Paper No. 8, NCEE-856.

Nayanananda, O.K. 2007. The Study of Economic Significance of Coastal Region of Sri Lanka in the Context of Environmental Changes of Pre and Post Tsunami. Coast Conservation Department and The Ministry of Environment and Natural Resources, 68 p.

New Civil Engineer, 2005. Sri Lanka tsunami repairs costed at £530M. Retrieved December 05, 2016, from https://www.newcivilengineer.com/sri-lanka-tsunami-repairs-costed-at-530m/535437.article

Ratnasooriya, H. A., Samarawickrama, S. P., & Imamura, F. (2007). Post tsunami recovery process in Sri Lanka. Journal of Natural Disaster Science, 29(1), 21-28.

Selvam, V., Ravishankar, T., Karunagaran, V.M., Ramasubramanian, R., Eganathan, P., Parida, A.K. 2005. Toolkit for establishing coastal bioshield. MS Swaminathan Research Foundation Centre for Research on Sustainable Agriculture and Rural development, Chennai, P 117.

Tanaka, N., 2009. Vegetation bioshields for tsunami mitigation: review of effectiveness, limitations, construction, and sustainable management. Landscape and Ecological Engineering, 5(1): 71-79.

United Nations, 2007. "The Human Toll". UN Office of the Envoy for Tsunami Recovery. http://reliefweb.int/report/sri-lanka/south-asia-tsunami-recovery-responsibilities-and-challenges. Accessed 15 June 2017.

UNDP, 2007. Sea Level Rises. 125-139.http://www.dmc.gov.lk/hazard/hazard/Report/UNDP%20BOOK%20CHAP%2007_%20SEA%20LEVEL%20RISE.pdf. Accessed 5 January 2018.

UNESCO, 2009. Tsunami risk assessment and mitigation for the Indian Ocean. Knowing your tsunami risk – and what to do about it. Intergovernmental Oceanographic Commission, Manuals and Guides 52.

Weerakkody, U., 1996. Impact of sea level rise on beach erosion and inundation of wetlands in the southwest coast of Sri Lanka. In Regional Workshop on Climate Change Vulnerability and Adaptations in Asia and the Pacific, Manila (Philippines), 15-19 Jan 1996.

Weier, J. and Herring, D. 2000. NASA. NASA, Accessed on 11th Dec. 2016.

Wijetunge, J.J. 2006. Department Tsunami on 26 December 2004: Spatial distribution of tsunami height and the extent of inundation in Sri Lanka. Science of Tsunami Hazards 24(3): 225-239.

Supplementary Information

Table S1. Land-use/cover change in acres between 2004–2005, 2005–2016, and 2004–2016.

2004 – 2005	Decreased	Unchanged	Increased
Dense forest	9.79	5126.42	0
Open forest	0	20 188.53	0
Garden	711.22	19 637.88	0
Other land	906.26	97 036.28	805.07
Paddy	557.32	47 442.97	24.91
Undefined	0	49.37	0

2005 – 2016	Decreased	Unchanged	Increased
Dense forest	0	49.37	0
Open forest	0	5136.20	0
Garden	0.44	20 348.65	0
Other land	15.79	99 320.05	23.13
Paddy	0	48 024.09	1.11
Undefined	0	20 188.53	0

2004 – 2016	Decreased	Unchanged	Increased
Dense forest	0	5136.20	0
Open forest	0	49.37	0
Garden	168.80	100 074.19	37.81
Other land	2.22	48 022.76	0.22
Paddy	106.97	20 242.12	0
Undefined	0	20 188.53	0

Made in United States
Orlando, FL
04 November 2026

90205473R00033